Pekin Robins
and Small Softbills

Pekin Robins
and Small Softbills

MANAGEMENT AND BREEDING

Peter Karsten

ISBN-13 978-0-88839-606-8
ISBN-10 0-88839-606-6

Copyright © 2007 Peter Karsten

Cataloging in Publication Data

Karsten, Peter, 1937 –
    Pekin Robins and small softbills : management and breeding / Peter Karsten.

    Includes bibliographic references and index.
    ISBN 0-88839-606-6

    1. Pekin robin. 2. Softbills. I. Title.

SF473.S64K37 2006   636.6'8   C2006-902277-1

All rights reserved. No part of this publication may be reproduced, stored in a retrieval system or transmitted, in any form or by any means, electronic, mechanical, photocopying, recording, or otherwise, without the prior written permission of Hancock House Publishers.

Printed in China — SINOBLE

Text, illustrations and photography by Peter Karsten

Technical editing support: Myles Lamont
Editing: Theresa Laviolette
Production and cover design: Ingrid Luters
Image editing: Laura Michaels

Published simultaneously in Canada and the United States by

HANCOCK HOUSE PUBLISHERS LTD.
19313 Zero Avenue, Surrey, B.C. Canada V3S 9R9
(604) 538-1114 Fax (604) 538-2262
HANCOCK HOUSE PUBLISHERS
1431 Harrison Avenue, Blaine, WA U.S.A. 98230-5005
(604) 538-1114 Fax (604) 538-2262

Website: www.hancockhouse.com
Email: sales@hancockhouse.com

Contents

How to Use This Book. 7
Key to Cover Illustration. 7
Acknowledgments. 8
Foreword 9
Preface 11

1 Introduction 13

2 Biology. 27
2.1   Appearance 27
2.2   Taxonomy 27
2.3   Habitat and Climate in the Wild 28
2.4   Adaptation and Behavior 29
2.5   Psychological Needs 33
2.6   Determination of Sex 33
2.7   Song and Vocalization 34
2.8   The Molt. 37

3 Housing. 39
     3.1   The Enclosure 39
3.2   Small Support Enclosures 41
     3.2.1   Portable holding cage. 42
     3.2.2   Hospital cage. 43
     3.2.3   Shipping box 45
3.3   The Breeding Aviary 47
     3.3.1   Size and configuration. 48
     3.3.2   The floor. 49
     3.3.3   The walls 50
     3.3.4   The roof 52
     3.3.5   Safety porch. 54
     3.3.6   Shift boxes 54
     3.3.7   Lay-out plans. 55
     3.3.8   Feeding stations 58
     3.3.9   Birdbaths 59
     3.3.10   Perches 59
     3.3.11   Positioning the aviary 60
     3.3.12   Winter housing 61
     3.3.13   Running water. 61
     3.3.14   Lighting 62
     3.3.15   Live plants 63
     3.3.16   Suitable Aviary Plants 64

4 General Care. 67
4.1   Capturing. 67
4.2   Trapping and Handling. 68
4.3   Predator Management. 70
     4.3.1   Rodents. 70
     4.3.2   Trapping rodents 71
     4.3.3   Other small predators 72
     4.3.4   Cats and raccoons. 73
     4.3.5   Birds of prey 74
4.4   Flock Management. 75
4.5   Observation 76

5 Acquisition, Quarantine, and Acclimation. 79
5.1   Acquisition. 79
5.2   Importation 80
5.3   Obtaining Background Information . . . 81
5.4   Judging a Bird's Condition 81
     5.4.1   Plumage 82
     5.4.2   Feet and legs 82
     5.4.3   Disposition and body condition. 83
     5.4.4   Eyes, nares, and beak. 84
5.5   Supplier's Facility. 84
5.6   Transport Home 84
5.7   Quarantine. 85
5.8   Acclimation 87
5.9   Release to Aviary 89

6 Breeding 91
6.1   Pair Behavior. 91
6.2   Breeding Season and Reproductive Age 93
6.3   Introduction of Mates 95
6.4   Preparation for Breeding. 96
6.5   Breeding Territory 97
6.6   Nesting Material. 100
6.7   Nest Supports 101
6.8   Mating and Egg Laying. 102
6.9   Incubation and Hatching. 103
6.10   Feeding Chicks 104
6.11   Early Loss of Chicks. 106
6.12   Fledging 108
6.13   Weaning 110
6.14   Banding 111
6.15   Record Keeping. 113
6.16   Breeding Events Calendar. 113
6.17   Breeding Statistics 114

7 Hand-rearing ... 117
7.1 Introduction ... 117
7.2 Basic Equipment ... 118
7.3 Brooder Construction ... 120
7.4 Hand Feeding Chicks ... 120
7.5 Fledging Hand-reared Chicks ... 122
7.6 Weight Monitoring ... 123
7.7 Behavior Considerations ... 124
7.8 Hand Feeding Log ... 125
7.9 Addendum: Protocol for Rearing Chicks from Day One ... 126

8 Feeding ... 129
8.1 Water ... 130
8.2 Protein ... 130
8.3 Carbohydrates ... 131
8.4 Lipids ... 131
8.5 Vitamins ... 132
8.6 Minerals and Trace Elements ... 134
8.7 Food Presentation ... 135
8.8 Feeding Utensils ... 136
8.9 Drinkers ... 137
8.10 Feeding Routine ... 137
8.11 Diet Information ... 138
8.12 Egg Cake ... 140
8.13 Food Value of Insects ... 141
8.14 Color Food ... 142
8.15 Fruits and Vegetables ... 143

9 Cultivation of Live Food ... 145
9.1 Insect Breeding Cabinet ... 147
9.2 Cricket Culture ... 148
   9.2.1 Cricket-rearing containers ... 148
   9.2.2 Hatching units ... 149
   9.2.3 Starting a culture ... 150
   9.2.4 Egg-laying and incubation ... 150
   9.2.5 Cleaning and handling ... 152
   9.2.6 Large-scale propagation ... 153
9.3 Waxworm Culture ... 154
   9.3.1 Breeding cycle ... 154
   9.3.2 Waxworm culture cabinets ... 155
   9.3.3 Hatching containers ... 155
   9.3.4 Rearing containers ... 156
   9.3.5 Food Medium ... 156
   9.3.6 Gut-loading ... 157
   9.3.7 Handling Moths ... 157
9.4 Mealworm Culture ... 158
   9.4.1 Starting, maintaining a culture ... 159
   9.4.2 Mite infestation ... 160
9.5 Zophobas (Superworms) ... 160
9.6 Lesser Mealworms (Buffalo Worms) ... 161
9.7 Fly Cultures ... 162
9.8 Whiteworms and Earthworms ... 165
   9.8.1 Whiteworm culture ... 165
   9.8.2 Earthworm culture ... 166
9.9 Collecting Wild Insects ... 166
9.10 UV Insect Trap ... 167

10 Conservation Breeding ... 169
10.1 Pekin Robins in Zoological Gardens ... 170
10.2 International Studbooks ... 171
10.3 Studbooks for Pekin Robins ... 172
10.4 Breeding Loans ... 172

11 Health Care ... 175
11.1 Introduction ... 175
11.2 Common Disorders ... 177
   11.2.1 Respiratory disorders ... 177
   11.2.2 Gastrointestinal disorders ... 179
   11.2.3 Sinus and eye disorders ... 179
   11.2.4 Scaly legs and swollen feet ... 180
   11.2.5 Fractures ... 182
   11.2.6 Head trauma ... 183
   11.2.7 Egg binding ... 183
   11.2.8 Rickets and related disorders ... 184
   11.2.9 Bleeding ... 185
11.3 Parasites ... 186
   11.3.1 Detecting the presence of parasites ... 189
11.4 Exertion Myopathy ... 191
11.5 Stress ... 192
11.6 Starvation ... 193
11.7 Weather Exposure ... 193

12 Other Species of Softbills ... 195
12.1 Introduction ... 195
12.2 A Palette of Softbills (species list) ... 197
12.3 Silver-eared Mesia ... 200
12.4 Minlas ... 203
12.5 Yuhinas ... 204
12.6 Laughing Thrushes ... 207
12.7 Bearded Tit or Reedling ... 208
12.8 Long-tailed Tits ... 211
12.9 Eurasian Tits ... 212
12.10 Bulbuls ... 215
12.11 White-eyes (Zosterops) ... 216
12.12 Eurasian Thrushes ... 219
12.13 Eurasion Robins, Bluethroats, Redstarts ... 220
12.14 Shama, Oriental Magpie Robin ... 223
12.15 Old World Flycatchers ... 224
12.16 Leafbirds, Fairy Bluebird ... 227
12.17 Tanagers, Euphonias, Chlorophonias ... 228
12.18 Honeycreepers ... 231

Glossary ... 232
Bibliography ... 239
Species Index ... 241
General Index ... 246

How to Use This Book

The approximately 100 topics discussed in this book are organized by a decimal numbering index. These decimal numbers indicate cross-reference of various topics within the text. Topics often have relevant expanded information in another chapter, and the decimal numbers indicate the chapter and subsections where this information can be found.

The reader may find that some key points of information are repeated in other chapters; while this repetition may lend emphasis, it is actually done to allow the chapters to be stand-alone essays. This enables the author, or readers, to relay condensed and comprehensive information on a particular subject to another aviculturist, thereby further serving the book's purpose, which is to serve as a resource tool to advance the breeding pekin robins and other softbills.

Key to cover illustration

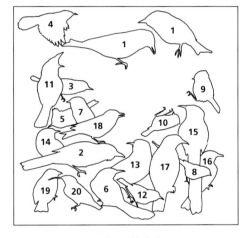

1. pekin robin
   Leiothrix lutea
2. silver-eared mesia
   Leiothrix argentauris
3. blue-winged minla
   Minla cyanouroptera
4. red-tailed minla
   Minla ignotincta
5. whiskered yuhina
   Yuhina flavicolis
6. yellow-throated laughing thrush
   Garrulax galbanus
7. bearded tit
   Panurus biarmicus
8. long-tailed tit
   Aegithalos caudatus europaeus
9. blue tit
   Parus caeruleus
10. coal tit
    Parus ater
11. red-eared bulbul
    Pycnonotus jocosus
12. Oriental white-eye
    Zosterops palpebrosa
13. dama
    Zoothera citrina
14. European robin
    Erithacus rubecula
15. white-rumped shama
    Copsychus malabaricus
16. verditer flycatcher
    Muscicapa thalassina
17. golden-fronted leafbird
    Chloropsis aurifrons
18. paradise tanager
    Tangara chilensis
19. yellow-collared chlorophonia
    Chlorophonia flavirostris
20. purple honeycreeper
    Cyanerpes caeruleus

Measures and abbreviations used in this book

Measures
The metric measures shall prevail over others for accuracy or clarity in this book. Metric measures are used in nutrition, medicine and other sciences. US and imperial measures are not always equal and the following conversion is used in this book:

Weight Measure Equivalents
1 ounce = 28.35 grams (g); US 1 ounce = 28.57grams
1 pound = 453.592 g = 0. 4359 kilograms (kg)
35.274 ounces = 1 kg = 1000 g

Fluid Measure Equivalents
1 US fluid once = 29.573 cubic centimeters (cc) or milliliters (ml)
1 US gallon = 3.785 liters (l)

Distance Measure Equivalents
1 inch = 2.540 centimeters = 25.40 millimeters (mm)
1 foot = 0.3048 meter (m) = 30.48 centimeter (cm)

Metric Units
1 kg = 1000 g
1g = 1000 milligrams (mg) = 1,000,000 micrograms (μg);
    equals 1 part per million (ppm)
1 liter (l) = 1000 ml or 1000 cc
1m = 100 cm = 1000 mm = 1,000,000 μm

Commonly Used Abbreviations

With scientific names:
sp. = species singular
spp.= species plural
ssp. = subspecies singular

Specimens in zoological inventories:
1.0 = one male
0.1 = one female
0.0.1 = one with sex unknown
Example: 1.2.4 = one male, 2 females, and 4 sex unknown.

Taxonomy
The taxonomy used in this text will be that of Sibley and Monroe's *Distribution and Taxonomy of Birds of the World*, 1990, and its 1993 supplement.

Acknowledgments

I wish to thank the many friends and people who graciously shared their experience and knowledge with me about caring for softbill birds, which led to writing this book.

Our late family friend Dr. Mervin McArthur who provided me with the first breeding hen and later his beloved pair of pekin robins, before he passed away, helped set the foundation with his birds and his great love and admiration for pekin robins. Mike and Elaine Manley, Grant Rishman, and the late George Bligh helped shed light on the subject of breeding pekin robins and mesias. Theo Pagel, Curator, Cologne Zoo, reviewed the manuscript and shared his experience and knowledge in aviculture, as did David Holmes, as far away as Howlong, Australia. Alan Lieberman, and Cyndi Kuehler of the Hawaii Endangered Bird Conservation Program and San Diego Zoo provided valuable insight into hand-rearing. Dr. Mark Finke and Dr. Ellen Dierenfeld reviewed and added to the chapters relating to nutrition and insect breeding by sharing their extensive knowledge of the science of entomology, nutrition, and especially wild animal nutrition. I am grateful to Dr. Chris Sheppard, Curator of Birds, Bronx Zoo, Wildlife Conservation Society, for the review of the manuscript. David Bender — a born aviculturist if there ever was one — passed on experience-tested advice, took on the task of establishing a Canadian studbook for the genus *Leiothrix*, and nudged me into writing this book, and then assisted with the technical edit. This was done with the encouragement of David Hancock, a highly accomplished aviculturist himself, who publishes specialty books on aviculture and wildlife with the primary objective of advancing and preserving knowledge in the field. I feel very much honoured by my mentor in wildlife conservation Dr. William G. Conway, Senior Conservationist, Wildlife Conservation Society, for writing the Foreword. Our friendship began nearly thirty years ago while we were directing zoological gardens entrusted to us in New York and Calgary. A further thank you is extended to my daughter Dr. Karen Karsten, who reviewed the section on Health Care, and my son Werner who helped me build my large walk-through aviaries. Last but not least, I thank my understanding wife, Margrit, who has maintained organization in our home, which has indeed been a real zoo for over forty years. Margrit coached my budding computer skills and calmed me down when I became frustrated by the bewildering cyber world.

Foreword

This delightful book by scientist, artist, and aviculturist, Peter Karsten, is an "owner's manual" for anyone keeping small insectivorous and frugivorous birds, the so-called softbills, which are delicate species almost never consistently bred in aviaries. Karsten's handsome, and astonishingly numerous, drawings and paintings, his meticulous instructions, and fascinating case histories provide a tested methodology for both care and propagation. His book breaks new ground in aviculture — beautifully — for both hobbyists and conservationists.

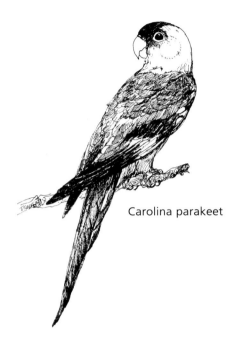

Carolina parakeet

Humans have kept wild birds as pets for thousands of years and Greek aviaries exhibited a wide variety by the fifth century b.c. The Greeks knew talking parrots seventeen years before Aristotle was born. Outside of zoos, however, most "cage birds" have been seedeaters that can be easily fed, such as parrots and finches. Today, poets and plumbers, financiers, and conservationists are keeping scores of delicate softbills, such as pekin robins and silver-eared mesias. Some aviculturists do so to enjoy their beauty and the closeness to nature they bring, others to resolve scientific puzzles, while still others are trying desperately to breed them to save them from extinction.

Of Earth's 10,000 species of birds, 1,213 are now considered in danger of extinction and the live-bird trade is among their dangers in some countries. Karsten makes it clear that the only way for aviculturists to sustain their collections is through conscientious attention to conservation concerns, as well as to care and collaborative programs of propagation. They must nourish viable populations and generations, not simply pairs of birds.

Peter Karsten was born in Göttingen, West Germany, immigrated to Canada in 1962 and soon began working at the Calgary Zoo. In 1974, he became director, skillfully modernized the zoo with new habitat exhibits, and produced a cascade of technical papers on zoo animal husbandry. Eventually, he was elected president of the Canadian, American and World Zoo and Aquarium Association. Upon retiring from Calgary Zoo in 1994, he devoted himself to wildlife painting, international zoo consulting work, and to breed birds.

Passenger pigeon

No one else has addressed the care and propagation of small insectivorous and frugivorous birds in such rich and loving detail. Karsten's efforts will help not only those who keep birds to improve their skills but, more especially, the birds themselves. While reading this unusual work, I could not help but think: If only such a book had been available for the Carolina parakeet and the passenger pigeon.

— **WILLIAM CONWAY**
Senior Conservationist and former President,
Wildlife Conservation Society, New York

*I dedicate this book to my wonderful mother, Marie Luise Karsten,
who celebrated her 100th birthday in May 2006
and still continues to foster my love for animals and nature,
and thus has made me very wealthy — forever.*

Preface

At the end of a demanding day in the office, I walked by an aviary at the Calgary Zoo and heard a pekin robin sing with exhilaration and seemingly great joy in its heart. It was still winter. The bird's spirit, anticipating spring, touched me and I listened to it with deep appreciation. I knew then that someday I would have my own pair of pekin robins.

A few more years went by until I retired, but soon afterwards I created an East Asian garden to house a pair of pekin robins. With the aviary built, I was ready to bring the birds to my Gulf Island home — which has a climate very similar to their home range — but none were available in the specialized pet shop.

Two years prior, in 1997, the Convention on International Trade in Endangered Species (CITES) had listed pekin robins in Appendix II. This meant they could no longer be imported for commercial trade.

I began to search bird clubs and the aviculture network and finally found one pair a day's travel away, but the hen died the next day, due to stress and a heavy infestation with gapeworms *Syngamus* sp. (chapter 11.3). Three months of intense searching went by before I found another hen. I realized that I would only be able to enjoy the pekin robin's song and delightful company in the future if I bred them myself.

Fortunately I could give the pair my almost undivided attention and they rewarded me with three chicks within two months. "Beginners luck" was the response by my avicultural contacts. To repeat and manage the luck became the new quest.

Close behavior observations to identify their needs, or to "get inside them" as we say in zoo circles, became the key approach. Keeping detailed daily records, studying relevant literature and exchanging knowledge with other aviculturists made luck more attainable. At the writing of this book, eighty-five chicks have been raised in seven breeding seasons from sixteen unrelated founders with second and third generation reproduction.

The overall purpose of the book is to promote the conservation of the species through the sharing of experience and information about proven methods to care for pekin robins and other softbills. It is designed to assist a potential new owner in assessing and advancing his or her capabilities to care for and breed these birds. It is a helpful reference book for me to pass along with birds that will be relocated to new owners to expand the breeding consortium.

Facing page: Male pekin robin singing.

The following chapters explore the demands on the keeper and breeder with a view to building cooperative breeding programs. The husbandry regimes and information presented in this book are based on my experience managing breeding aviaries on the West Coast of British Columbia, Canada. Other locations have different climates, which must be taken into consideration.

Another important focus of this book is the on-site production of live food. There is great comfort in having sufficient quantity and quality of the required food insects when breeding pairs begin to nest and raise their chicks. Through various trials, simpler methods were discovered for the successful propagation of food insects. This section on its own is of interest to animal keepers who are in need of live food for their birds, reptiles, amphibians, and small mammals.

The book is equally applicable for zoological gardens and public aviaries. I know from experience that public facilities must show diverse species of birds and other animals for educational and recreational reasons. Often reproduction is not consistent or even possible for many species. It is in this area where private breeders and zoological gardens can work well together, by dividing the roles between intense off-exhibit breeding and exhibition of specimens that are not required for the breeding program. Many zoological facilities are expanding their own breeding programs for softbills, but facility space remains at a premium and, if it is available, it is usually reserved for species which are unlikely to be obtained by a private person.

A reader who is contemplating the acquisition of a pekin robin or other softbill should first read this and/or other relevant literature to appreciate the extent of the commitment and care requirements, and only then make an informed decision. All too often valuable birds have been purchased before research was done or the proper enclosure prepared, resulting in disappointment.

Gaining experience and gathering insights continues forever; finally making the decision to finalize the manuscript was a hard call. I welcome communication with other aviculturalists or bird enthusiasts in the spirit of sharing experience and increasing collective knowledge in this field.

It is a privilege to pass on my experience to the reader, and to help expand the stewardship of pekin robins and other softbill species for their conservation.

— PETER KARSTEN
January 2006

1

Introduction

The tradition of keeping wild birds as companion animals is as old as human society itself. Numerous and varied human-bird relationships have developed over the centuries, ranging from maintaining a living food supply in the form of domestic poultry flocks used for the production of meat or eggs, to the ancient but still much-practiced art of falconry, to the beloved companion bird or family pet, to the exhibition of exotic bird species in public zoos or private aviaries, to endangered species recovery and propagation programs, and many others.

Some species of birds can readily be accommodated in the environment of an average household, and have become highly domesticated and totally adapted to the conditions of captive care and breeding. Other species have not been fully domesticated, but are commonly kept and bred in various aviaries. These birds give an immeasurable amount of joy and decidedly enhance the quality of life of their keepers, particularly the urban pet owner who has little opportunity to stay in close touch with nature. The focus of this book is on the keeping and breeding of the pekin robin *Leiothrix lutea* and similar species of small, softbill birds in private or public aviaries. Other names for this bird are red-billed leiothrix, Chinese or Japanese nightingale, and hill tits.[1]

Figure 1.0-1 The Gibber canary is an example of a highly domesticated fancy breed of the wild canary *Serinus canarinus*.

Softbills are often considered too difficult and demanding to fit into the daily life of a busy person or family. Granted, the more popular species of seedeaters are easier to care for simply because the dry food they eat does not spoil easily. Seedeaters can be provided with a self-feeder containing a sufficient supply of food for several days in an emergency, or to bridge times of other pressing commitments. A softbill, and more to the point, the food presentation during the nesting season, requires daily attention. Consistent breeding and rearing success can only be achieved with a regular supply of live insects during the nestling period. The rewards for

---

1. Christine Sheppard, pers. com. 2006.

accepting the challenge include not only the unmatched, melodious and beautiful songs of softbill birds, but also their high spirits.

This book may also serve as a guide for the reader to decide whether a softbill in general, and a pekin robin in particular, is indeed the right bird to acquire.

Pekin robins, once imported by the thousands, are no longer readily available since the international, commercial trade in wild-caught birds of this species is prohibited under endangered species legislation.

If this delightful and popular softbill is to be preserved in human care for the enjoyment of future generations of bird-enthusiasts, it is incumbent upon the aviculturists to work together in cooperative breeding programs, including studbooks and careful genetic management of the available founders and offspring, to ensure that healthy, self-sustaining populations are maintained for the long term (chapter 10.3 and 10.4). For the time being, it is urgent to place potentially reproductive pekin robins held in our aviaries into the dedicated hands of qualified aviculturists until the species is propagated in sufficient numbers to supply birds for other interests.

The proper care for any kind of animal requires a good understanding of its structure (anatomy), its bodily functions (physiology), its mind and behavior (psychology and ethology). The genetic make-up creates typical form and function; beyond this, individual traits emerge through encountered experiences and specific environments. Hence, a wild-caught bird is in many ways a different animal than its aviary-reared counterpart, and a parent-raised bird is different again from a hand-raised bird. They may all look alike and eat the same diet, but often require somewhat different husbandry.

This point is made to recognize that each bird is unique. Many decisions on how to care for a bird can only be made on location. A proven avicultural technique that fits one situation may not fit another. The secret to success and fulfillment in one's relationship with birds and other animals is to recognize them as individuals and strike a form of "partnership" with them.

Hopefully, this book will remove some of the mystery from keeping and breeding pekin robins, and help to widen the circle of softbill breeders.

While the keeping of these birds offers its own rewards, the satisfaction of being able to pass a viable, self-sustaining, *ex situ* population of these birds to the next generation of aviculturists is an even greater reward. This is the mission of the author and other like-minded breeders.

Chapter 1 ■ INTRODUCTION

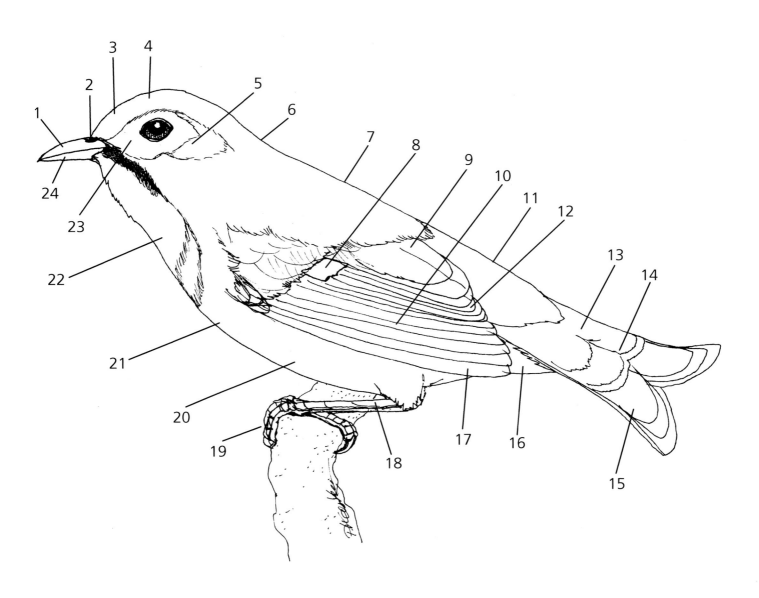

Figure 1.0-2 The topography of a pekin robin.

1 upper mandible
2 nares
3 forehead
4 crown
5 ear coverts (auriculars)
6 nape
7 back
8 small wing patch
9 tertiaries
10 inner primaries
11 rump
12 secondaries
13 upper tail coverts
14 extended upper tail coverts
15 tail feathers
16 under tail coverts
17 outer primaries
18 legs
19 feet
20 belly
21 breast
22 throat
23 facial patch
24 lower mandible

Phenotypes of four subspecies of pekin robins

[*See* chapter 2.2]

Key to color plate 1.0-1 facing page

Red-billed leiothrix *Leiothrix lutea lutea*:
    1 – facial patch: white/gray
    2 & 3 – crown and nape: olive green wash
    4 – back: gray
    5 – rectangular small wing patch: yellow
    6 – primaries at base: orange-red
    7 & 8 – inner and outer primaries: yellow margin
    9 – upper tail coverts: gray
    10 – under tail coverts: white
    11 – legs and feet: yellow
    12 – throat and breast: vivid orange and yellow.

Indian red-billed leiothrix *Leiothrix lutea calipyga*:
    1 – facial patch: yellow wash
    2 – crown: with yellow tinge
    3 & 4 – nape, side of head and back: olive green
    5 – small wing patch: yellow
    6 – primaries at base: red
    7 – inner primaries: orange margins
    8 – outer primaries: yellow edge
    9 – upper tail coverts: olive gray
    10 – under tail coverts: yellow wash
    11 – legs and feet: yellow brown
    12 – throat and breast: yellow and pale orange
Smaller than *L. l. kwangtungensis*

Kwangtung red-billed leiothrix *Leiothrix lutea kwangtungensis*:
    1 – face: yellow wash
    2 & 3 – crown and nape: golden olive
    4 – back: soft olive gray
    5 – wing patch: orange and short
    6 – base of primaries: rich chestnut brown
    7 – inner primaries: more yellow than *L. l. calipyga*
    12 – throat and breast: deep yellow with faint orange tinge
Largest subspecies.

Yunnan red-billed leiothrix *Leiothrix lutea yunnanensis*:
    1 – facial patch: white/gray
    2 – crown: dull olive
    3 & 4 – nape and back: gray
    5 – small wing patch: yellow-orange
    6 – base of primaries: orange
    7 – inner primaries: no yellow on edge and are black
    8 – outer primaries: yellow crimson edge
    12 – breast: strong reddish orange
Generally the plumage is less vivid than *L. l. calipyga* and *L. l. kwangtungensis*.

Chapter 1 ■ INTRODUCTION

Color plate 1.0-1

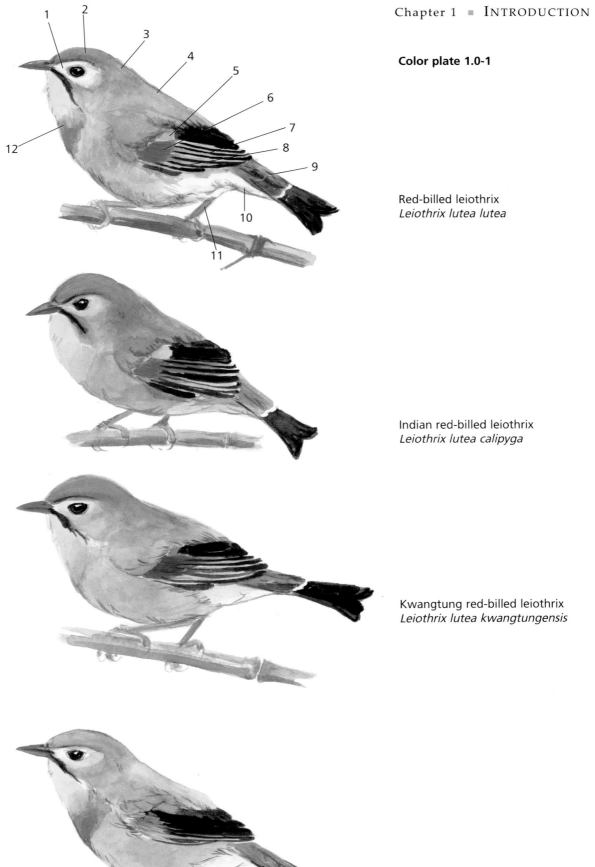

Red-billed leiothrix
Leiothrix lutea lutea

Indian red-billed leiothrix
Leiothrix lutea calipyga

Kwangtung red-billed leiothrix
Leiothrix lutea kwangtungensis

Yunnan red-billed leiothrix
Leiothrix lutea yunnanensis

Biology and Housing

1. Male pekin robin showing dark coloration at base of bill in winter season.

2. The author's first pair of pekin robins (1999) female on left, male on right.

3. Pekin robins have bright red bills during the breeding season.

4. All tail feathers may be dropped at once during the molt. Primaries are replaced in stages to maintain ability of flight.

5. Female pekin robin in a bamboo grove. Note small facial patch around the eyes.

6. Pekins form same-sex buddy relationships. Theses are two males.

7. A park in Hanghzou, PRC. Pekin robin habitat has dense vegetation cover.

Biology and Housing

1. Juvenile sliver-eared mesias and a pekin robin roost together. Adults will form inter-specific buddy relationships.

2. A well-planted aviary is a prerequisite to successful breeding.

3. A large walk-through aviary allows the addition of other compatible species to the breeding territory of a pair of pekin robins.

4. The author's early breeding aviaries grew by "evolution."

5. "Laddered" undergrowth is critical for fledging chicks to find a safe roost.

Housing and General Care

1. Feeding station showing pullout tray and contamination shield.

2. Mice are easily trapped in the partially pulled out food tray.

3. A garter snake searching for tree frogs could mistake newly hatched chicks for a meal.

4. A centrally hooked up water dish with in and out feed water lines.

Breeding

5. Male in a dominance pose, which is also used in courting.

6. Males showing substantial scars from a serious earlier territorial dispute.

7. Nest supports are made of wire wrapped with raffia.

8. The nest support is put in place just as the nest is started.

Breeding

1. Nests are secluded in dense vegetation. Usually three eggs are laid.

2. Eggs (left to right): Silver-eared mesia, pekin robin, red-tailed minla, and Oriental white-eye.

3. Pekin robins one day old.

4. Seven days old, ready for closed-banding.

5. As chicks mature the parents bring several insects to the nest at a time.

6. Male pekin robin at the nest.

7. The author's first pekin chicks, three days after fledging.

8. The fledged chicks try to roost together.

9. The last sibling hatched may be smaller, but usually catches up in time.

Hand Rearing

1. A chick is moved into a hospital cage due to imminent fledging.

2. Silver-eared mesia chicks do a lot of preening just before fledging.

3. A chick learning to recognize food by feeding out of a perch-feeder.

4. Weight monitoring is important to measure progress.

5. Pekin robins roost together at night and disguise their shape.

Cultivation of Live Food

7. A converted deep freezer chest is ideal to raise crickets. Note thermostat on left and two heat lamps under the lid. Floral foam with eggs elevated on shelf and pail, crickets, and egg cartons on floor.

8. View into insect breeding cabinet.

9. A converted refrigerator for mealworm cultures.

10. House crickets, female with ovipositor on left, male on right. Inset: shedding nymph.

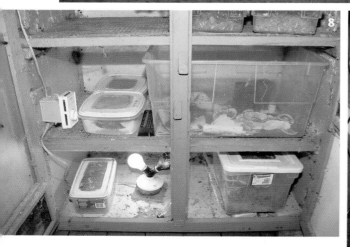

Cultivation of Live Food

1. Crickets laying eggs into floral foam block.
2. Large number of nymphs hatching from foam block.
3. Greater wax moth larva, cocoon, pupa, moths.
4. Egg strips on the rim under the lid of a rearing container.
5. Mature waxworm in the medium and under the lid for pupation.
6. Beetles of giant mealworm, mealworm, and lesser mealworm.
7. Larvae and pupae of giant mealworm, mealworm, and lesser mealworm.
8. Mature mealworms migrating to folds of brown paper for pupation.
9. Lesser mealworm trap and sifting equipment. Beetles and larvae in trap dish.
10. Homemade UV insect trap.

Healthcare

11. Pekin robin chick with rickets.
12. Severely egg-bound hen under heat lamp in the aviary.
13. Fledged chick attempting to find a secluded perch and prone to muscle burn-out.

Other Species

1. silver-eared mesia *Leiothrix argentauris*
2. blue-winged minla *Minla cyanouroptera*
3. red-tailed minla *Minla ignotincta*
4. blue tit *Parus caeruleus*
5. coal tit *Parus ater*
6. white-rumped shama *Copsychus malabaricus*
7. Oriental white-eye *Zosterops palpebrosa*
8. European robin *Erithacus rubecula*

Figure 1.0-3
European finches which have been housed or bred by the author with pekin robins and mesias in large, well-planted aviaries.

Top row: bullfinch *Pyrrhula pyrrhula*, greenfinch *Carduelis chloris*, linnet *Acanthis cannabina* **Center:** goldfinch *Carduelis carduelis* **Bottom:** siskin *Carduelis spinus*, redpoll *Acanthis flammea cabaret*, chaffinch *Fringilla coelebs*

Figure 2.0-1 Pekin robins in their homeland.

2

Biology

2.1 Appearance

The pekin robin is about sparrow size. Adult males usually weigh 3 to 5 grams more that females.Its average adult weight ranges from ¾ to 1 ounce (approximately 22 to 30 grams). The total body length, measured from the tip of the bill to the tip of the tail, is about six inches (15.2 cm). The slender bill is bright red during the breeding season, but the base of the beak changes to a dark charcoal color in the winter. The many subtle and contrasted vivid colors of the plumage make it a very attractive bird.

There is no obvious difference in the appearance of the sexes (dimorphism). The trained eye of a pekin robin enthusiast can see a difference in the facial marking and may get clues from the white line on the outer edge of the extended upper tail coverts, but it would be unwise to let the reader think sex can always be told by the plumage (see Determination of Sex, chapter 2.2.6).

Occasionally, hens have as much white on the edges as some males. Juveniles show more distinct markings after the first molt. Both sexes have the elegant outward curving tail feathers. Birds that have been handled or shipped often lose the extended tail coverts. Many softbills readily shed their tail feathers when seized by the tail (chapter 4.1). These particular feathers seem to shed relatively easy and are at times absent.

Sex can be determined by vocalization, and by DNA testing as mentioned later in the text (chapter 2.7).

Presumably the pekin robin got its name from the fact that many of the birds, exported via Peking, China, had a reddish patch on the chest reminiscent of the European robin *Erithacus rubecula* (chapter 12.13). Dictionaries define "pekin" as a striped silk fabric, and on pekin robins the distinct markings on the edges of the primaries do produce a striped pattern.

2.2 Taxonomy

Pekin robins are not true robins and thus not members of the sub-

Figure 2.1-1 Measuring body length.

Figure 2.1-2 Extended upper tail coverts of male, with the white line, (left) and female (right).

Figure 2.1-3
European robin.

Figure 2.2-1 Wrentit.

Figure 2.2-2 Eurasian bearded tit or reedling.

family Turdinae, the thrush family, which includes the European robin (chapter 12.13).

Taxonomically, pekin robins are classed as Aves (birds), belonging to the order Passeriformes, or perching birds, and further defined as members of the family Sylviidae, the subfamily Sylviinae, and finally the tribe Timaliini that has the common name "babblers" (chapter 12.3 to 12.7).

Babblers constitute one of the most diverse tribes in appearance, embracing many species. They are found in the Asian, African, and Australian regions with only one species in the Americas, the wrentit *Chamaea fasciata*, which lives on the West Coast of the United States, and one species which extends its range into central Europe, the Eurasian bearded tit *Panurus biarmicus*. Bearded tits (chapter 12.7) are kept and bred by aviculturalists; the wrentit is not. Other babblers which we find in aviculture are introduced in chapter 12.3 to 12.6.

The scientific name of the pekin robin, also called Chinese nightingale or red-billed leiothrix, is *Leiothrix lutea*. *Leiothrix* defines the genus and *lutea* (or *luteus* in some references) the species. A subspecies name may be added to it. There are several subspecies of pekin robins.

I have attempted to illustrate the differences between four of the subspecies from text references of museum specimens (color plate 1-1, page 17). The feather pigmentation of bird skins that are kept in museum storage can fade over time and complicate identification; hence the illustrations can only serve as a guide. The approximately 50 wild caught birds I have seen resemble the subspecies *L. l. lutea*.

Readers may see some variation from bird to bird. Diets and environmental conditions play a role in plumage color, as we will see in chapter 8 (Feeding).

2.3 Habitat and Climate in the Wild

In the wild, pekin robins range from the western region of the Himalayan Mountains of Kashmir through Tibet and Nepal to the southeast of Assam; northern Myanmar; northern Vietnam to Tonkin; in China from Sichuan in the north to the very south and the east. They can be seen in mountain forests up to almost 10,000 feet (approximately 3,000 m) in elevation, which explains their tolerance for temperate climates. Their preferred habitat is the underbrush of forests, plantations, and varied landscapes with patchy brush and tree cover. The species is relatively reclusive and not commonly observed in the open countryside.

The pekin robin is a favorite cage bird in Asia, where it is fre-

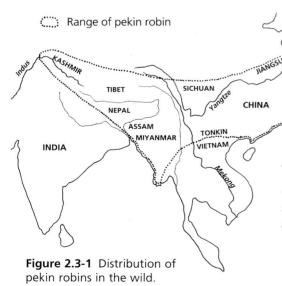

Figure 2.3-1 Distribution of pekin robins in the wild.

quently trapped for the bird market. In the past, thousands were exported to Europe and North America. Today we find free-living breeding populations in Hawaii, and on Gran Canaria and Tenerife, Canary Islands.

Gradual acclimation of once wild-caught birds was necessary, since one could not establish from where birds originated. It stands to reason that the southern lowland and island populations could not become accustomed to unheated outdoor aviaries as quickly as those from the northern, higher mountain ranges. Now that the species is protected under the Convention on International Trade of Endangered Species (CITES), the importation of wild-caught birds from their native range has effectively ended (chapter 10).

Acclimation to colder ambient temperature must still be done with domestically raised birds. It is important to investigate the circumstances of the originating aviary to match the climatic parameters (chapter 5.8). The relocation of birds from cold to hot climates requires acclimation as well.

2.4 Adaptation and Behavior

Species-specific adaptation (special or specific adaptation) and behavior allows the pekin robin to survive and thrive in a particular ecological setting, its niche. Finding sufficient food is fundamental to its survival strategy. Evolution has shaped the bills of birds for this; hence a good hint for their care is gained from the description "softbill" or "softbilled" bird. This does not mean they have soft and flexible bills, it means that they eat soft foods. In other languages they are called "soft-food-feeders", which is more to the point.

Seedeaters can shell hard seeds because of the shape of their bill, adaptation of their mandibles, and tongue; finches and parrots have this ability. Parrot-like birds (Psittacines) are often called hookbills by bird fanciers.

The shape of a bird's bill generally reveals the specialized method of food gathering. A cardinal, a seed-eating finch, has a bill or beak that is designed to crush and shell husked seeds. A pekin robin has a slender bill, which helps the bird in probing for insects and their eggs, and picking the flesh of fruit. A sunbird or honeycreeper (chapter 12.18) has a long and curved beak that allows it to reach deep inside a flower, or to pierce its base to reach its nectar; and the powerful beak of a parrot can dislodge and crack large seeds and nuts.

In aviculture softbills are grouped according to the type of their main food intake. Frugivores are fruit eaters, semivores are seed eaters, insectivores are insect eaters, nectivores are nectar eaters, and those who mix up the menu are called omnivores. But

Figure 2.4-1 The shape of bills reveals the food source.

Figure 2.4-2 Blue-breasted quail and hatched chicks.

Figure 2.4-3 Finches are "dish sitters" while feeding.

Figure 2.4-4 Shamas will take several food items, while Pekin robins take one at a time.

it is not that neat and tidy. Some softbills have carnivorous tendencies and many change diets between seasons from one food group to others. For example, during the breeding season hummingbirds feed insects to their chicks and so do most finches.

The important part to remember is that birds which are feeding chicks in the nest, and fledged young, must be provided with food high in animal or complete protein to ensure the chicks can develop at the fast rate they do (chapter 8.2).

Birds which have naked chicks at hatching time and brood and feed them in a nest, as pekin robins do, are called altricial or nidicolous. Those which leave the nest at hatching time, such as the gallinaceous birds including chickens, pheasants and quail, for example, are called precocial or nidifugous. The blue-breasted quail *Coturnix chinensis* is one of the smallest of the precocial species; the ostrich, to no one's surprise, is the largest.

Pekin robins and many other babblers are insectivores/frugivores with omnivorous tendencies. This makes them easy birds to keep on a maintenance diet, but it must be emphasized that raising chicks cannot be accomplished without plenty of live insect food (chapter 6.10). Pekin robins can be observed eating a few seeds, but their digestive system is not designed to extract nutrients in sufficient quantities to sustain them on husked seeds. On finch seeds alone a pekin robin would soon suffer malnutrition. Seedeaters shell the seeds and have a crop for pre-digestion, plus muscular stomachs with ingested grit to mechanically process seeds or "hard" foods. Pekin robins digest their food more like we do, chemically and much less mechanically (chapter 8).

Another point to consider is the way birds gather food. Seedeaters settle down at a dish of seeds or a seed-bearing patch of plants and fill their crop, to last them for an extended period of time. Pekin robins typically hunt throughout the day for insects and other food that is dispersed within their habitat. The resting periods are shorter to allow for enough time to find sufficient food. Insectivores are very active and agile, which has several implications for their husbandry.

Pekin robins usually dart to the food dish and pick up one item to be consumed away from the dish in the cover of the vegetation, or on a favored perch. If they are housed in an aviary with seedeaters they are clearly at a disadvantage to compete for the choice food items, because the seedeaters will arrive at the dish and remain there until they have eaten their fill. Pekin robins are reluctant to chase even smaller finches away and must do this for every re-approach.

This is one reason why pekin robins seldom successfully raise their young in a busy community aviary. The "dish sitters" will hog the mealworms and other insect food simply because of their feeding behavior. This can be countered to an extent, depending on the population mix and density, by setting up more than one feeding station. Shamas, (chapter 12.14) incidentally will stay at the food dish and take several insects before moving away (figure 2.4-4).

Babblers are very active birds by nature, even when food is provided and they have no need to search for it. They need opportunities to play out the inborn motor drive. If kept in small cages (chapter 3.1) the birds will develop stereotypical behavior patterns by relentlessly going through the same sequence of movements. In a small cage they will damage their otherwise splendid plumage and, in my experience, soon fall into a pitiful condition. This species has been kept as aviary birds for over one hundred years because of their hardiness, adaptability to human home environments, and their acceptance of a wide variety of food items. With proper care pekin robins can live more than twenty years under human care.

Pekin robins love to bathe. A shallow birdbath with dripping water is frequently visited for a thorough bath, even in cold weather. Shallow dishes with a drip feed and drain line are ideal. When the drip system is turned on, soon the entire flock engages in an infectious bathing party, which spreads to others in adjoining enclosures (chapter 3.3.9).

Some writers suggest running water is conducive to nesting. More to the point, it stands to reason that rich, moist biotopes with a great wealth of insect life may trigger readiness to nest. The presence of food supply is a stimulus to nesting. Well documented is the relationship between the numbers of eggs laid by a snowy owl and the number of lemmings the male presents to her at the nest.

Most intriguing is the ant-eating behavior. The bird will seize an ant in its beak and will abruptly and repeatedly twist around to touch its tail as if struck by fits. Closer observation gives me the impression that the bird is testing to see if the ant is still biting before it is swallowed.

"Anting" is described in many species by ornithologists. Some suspect the bird is using the formic acid ants release to deter parasites and condition their plumage; others believe the birds react to a skin sensation. I see pekin robins and mesias only touching their tails two to three times very quickly and then immediately consuming the ant. Only the large ¼ to ½ inch (6 to 12 mm) carpenter ants *Camponotus* spp. are treated this way; other ants and termites are eaten immediately.

Sleeping together on the perch is enough to convince anyone that a single bird is a deprived bird, when two will melt into one body of fluffed feathers with two tails protruding from the outline. One could assume that this silhouette confuses the perceptions of predators. It certainly conserves energy, apart from the

Figure 2.4-5 Head twisting is a form of stereotypical behavior.

Figure 2.4-6 Testing the biting ability of a carpenter ant (inset).

Figure 2.4-7 Two roosting birds disguise their shape to predators.

Figure 2.4-8 Sunbathing resembles a very sick bird.

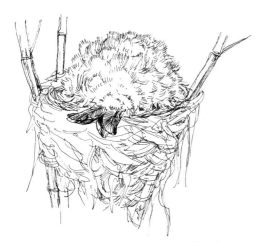

Figure 2.4-9 At night the hen ruffles her feathers to disguise herself on the nest.

more intangible notion of giving a sense of security and social contentment.

Sunbathing is a peculiar and terrifying behavior to the novice keeper. The bird will lay on its side with the beak wide open, eyes half closed, the wings spread away from the body, and absolutely motionless in an overstated display of death. All that is missing are the skyward-stretched feet. The idea is to let the sunrays reach the skin. Mutual preening, also called allopreening, is a delightful behavior to watch, and much easier to take.

Readers should also be warned of yet another behavior. If you check on the incubating or brooding hen at night — it is always the hen that takes that duty — you will see again what appears to be a dead bird; so dead that you may think decomposition has turned it into a bloated globe of loose, gray feathers. It even disguises the shape of the nest rim. This is pure mastery of deception. There is no sign of the colorful wing patterns, not to mention her head. The hen will do this also on very warm summer nights, proving that the behavior is a form of disguise and not energy conservation.

Being aware of these behaviors will spare you at least some heart-stopping encounters with perfectly happy pekin robins.

A pekin robin's memory of escape routes, and the deliberate flying directly towards your head past your ear to avoid the swooping net, is nothing short of remarkable. All too frequently a gap left between your wrist and the cage's doorframe is quickly detected and instead of rushing away from your hand the bird will dart towards it to escape.

But there is a helpful side to the coin. Should one escape, or even a pair together, they will explore their newly opened-up territory with their excited babbling; but while you think you will never see them again, they will, if left in peace, most likely return to their sleeping perch and allow you to close the door behind them. Pekin robins are very loyal to their sleeping spot and rarely shift to another. This helps if you have interconnected indoor and outdoor aviaries. If you bring the birds in often enough and force them to roost inside, they will eventually stick with the indoor roost and can be left to their own devices during the cold weather season. The level of seclusion (plant cover) must, however, be comparable otherwise they will change their mind again. Other very interesting behavior relating to feeding and chick rearing will be presented later (chapter 6.10).

Always remember that the pekin robins' curiosity is their addiction and one must always think ahead of them. For example, for safety reasons never set a mouse trap or leave a partly filled watering can where they could have access to it, or live electrical wiring, and so on.

2.5 Psychological Needs

Behavioral study has opened new doors to understand the psychological needs of animals and thus adds a new dimension of care — if not obligation — to animal husbandry. Pekin robins and their relatives are highly social birds. They need the company of their own kind, like the proverbial lovebirds, but not necessarily with birds of opposite sex. A single pekin robin will even strike a close friendship (buddy relationship) with another "contact" species for mutual grooming and roosting (figure 2.5-1). To deprive them of company causes them psychological hardship.

A pair, or a flock, of pekin robins "babbles" constant contact calls when they are traveling through dense bush and when out of sight of each other. A single female will call and a male sing incessantly to attract a mate or companion. When the species was marketed as a good caged songbird this behavior led to the exploitation of the species and exportation of almost exclusively males, because females do not sing.

The male has a contact song that is presented throughout the day and seasons in order to locate a companion. While for the uninformed bird lover this may appear desirable, it is rather selfish to keep a single pekin robin just for its song. The urge to be paired up commonly leads to the formation of single sex couples. Many such "mated pairs" of pekin robins have been sold over the years as true pairs. Mysteriously, to the disappointed owner who paid a premium for a bonded breeding pair, reproduction was hoped for, in vain.

Pekin robins and most softbills do not belong inside a small, traditional cage. In my view it is biologically incorrect, and outdated, for advanced aviculturists to keep these birds in such containers. Confined cages are recommended for special circumstances to acclimate, hospitalize, trap, or transport softbills, but not for their permanent home. Proper housing is explored in chapter 3.

2.6 Determination of Sex

The reliable determination of the gender of a pekin robin is, of course, necessary to establish a pair for breeding purposes. As noted above (2.1), sexual differences in the appearance (phenotype) are marginal at best. The male has generally more color and a more distinct whitish facial patch. The white edges on the two extended upper tail coverts are generally more prominent in the male; however, this is not reliable as discussed earlier. A pair that has lived in the same environment and on the same diet may show a relative difference in appearance. Unfamiliar birds, which are viewed for the first time, can be highly puzzling.

Social behavior is most misleading to determine the gender.

Figure 2.5-1 Single pekin robins and mesias will readily form "buddy" relationships.

Figure 2.5-2 Ruffling feathers on head and throat is an invitation to grooming.

Figure 2.6-1 The facial patch is usually larger and more distinct in the male.

Figure 2.6-2 A blood feather has serum and blood in the shaft, helpful in DNA sex analysis.

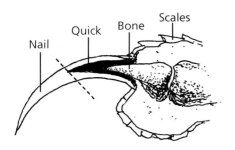

Figure 2.6-3 Position of cut to obtain blood for DNA testing.

Two birds of the same sex bond quickly and may even stay together if other birds of the opposite sex are introduced. Occasionally we need to break up a pair (siblings or same-sex "buddies") for breeding purposes. In that cases the bonded couple should be separated out of sight and sound from each other to be paired with new mates, otherwise, at best, new pair formation is delayed.

The birds can certainly be sexed by DNA testing which reveals genetic characteristics (genotype) and, along with it, the sex. It requires catching the bird and pulling out one or more feathers, which will have a few root cells attached. The larger feathers, such as tail or wing feathers are usually used, but smaller body feather are suitable as well. A large blood feather is ideal, one that is in the process of re-growth. It has enough blood and serum in its soft shaft to make the testing easy.

Another way is to clip a toenail shorter than one would normally do, to reach the blood vessel inside (quick). A drop of blood, when placed on a special card supplied by the laboratory, will be sufficient to make a conclusive test. If a lot of bleeding occurs, the bird should be kept in a holding cage until it has stopped before it is released to a big flight. At times the blood vessel will not seal and too much blood is lost, which can put the bird into trouble. In that case the nail has to be treated, as noted under Health Care (chapter 11.2.9).

A local veterinary clinic will be able to provide you with the address of a DNA testing lab.

2.7 Song and Vocalization

Fortunately, pekin robins can be sexed by their vocalization, which is rather diverse in order to serve a variety of purposes. The male has a form of signature (contact) song comprising about seven varied notes. The emphasis is on varied, because the female does not vary her notes. She has a monotonous, plaintive set of three to five notes, reminiscent of the call of a few days old poultry chick. A male is also capable of emitting this call, but he very rarely does. It may be heard when he is separated and frightened after being transferred to a confined shipping crate, for example.

I have sexed about 100 birds and paired them easily by vocalization without fail — so far. None of the hens ever sang. A male sings to attract a companion, to stake out his territory, or for the sheer joy of it. You may even hear him sing during the molt. Young males begin to practice when they are close to four weeks of age. It is quite heart-warming to hear them learning this skill.

Pekin robins have innate singing talents, but learn much of their repertoire from their parents and other pekin robins. They may also mimic other species. In addition, the male pekin robin has a more complex, territorial song and a softer courtship song. There are others calls: a scolding call, which is an uninterrupted loud series of harsh chattering sounds, and which often causes the whole flock to chime in; a short warning call, sounding like "chur-rp", that causes everyone to take cover; and the familiar constant uttering of soft single notes, especially when the birds explore new territory. The hen emits a soft, repeated whining call when she invites the male to mate. A breeder soon recognizes it as a welcomed sound.

To get the male to sing or the female to vocalize, more or less on command, they must be separated from other birds — out of sight but not sound. The male will try to make contact by using his signature song, unless he is stressed and very frightened. Other males will answer, and so will females with their call. By closely observing each bird, the sex is established, with the proviso that a male may still call like a hen in rare circumstances.

Birds that call like hens should be observed over several days. If a separated bird that is healthy and not under stress does not sing, and does not vocalize like a male but calls like a hen, it can be safely assumed that it is a hen.

Tape recordings work well to stimulate vocalization. I have sent recordings to people who had single birds, which helped them to establish the bird's sex. For the most part isolated males will sing immediately, even under some stress. I bought two pairs of pekin robins in 1999 and discovered within less than an hour that all were males. Conversely, I obtained two "males" from another fancier who had them for several years, only to discover that they were both hens.

The willingness to sing, even under duress, made pekin robins coveted songbirds in their homeland. They were, and still are, sold and kept in very small cages that are carried to a garden, park, or other public place to delight the owner and friends with their song.

Figure 2.7-1 Pekin robins are often kept for their singing in small cages in their country of origin.

This trait also made the pekin robin popular as a cage bird in Europe and the New World. Its eager singing, adaptability, and longevity in captivity led to the capture of thousands of pekin robins. Generally only males were kept or subsequently exported, which is one reason why captive breeding was uncommon. Besides this, the low cost and availability of pekin robins gave little incentive to make the effort. This changed dramatically after export and import for commercial purposes ended in 1997.

Once we understand what motivates a pekin robin to sing, we can hardly justify keeping a single pekin robin without a companion bird. This species is not as domesticated as the budgerigar, or as contact oriented, hence it does not make a good one-on-one pet.

Wild-caught birds have a species-specific song by which they can be identified. A bird raised in a mixed aviary without the proper teaching by its own kind will pick up quite a cocktail of songs. Hand-rearing chicks in isolation from audible songs of "real" pekin robin males may result in poor species-specific singers. A hand-reared chick can be trained by tape recordings. The preservation of the original song must be considered in breeding birds over several generations in *ex situ*.

At this time there are still males alive which were caught in the wild, and as long as the chicks are raised in a setting where these males teach them, their song will be passed on to future generations. It is believed that the chicks memorize the song while still in the nest. Hand-rearing chicks in isolation from audible songs of "real" pekin robin males may result in poor species-specific singers. Juveniles begin to practice their contact calls/songs when they are close to five weeks old, which coincides with the appearance of the first orange throat feathers (photo page 19, image 1 Biology and Housing).

Figure 2.7-2 Song imprinting begins in the nest.

2.8 The Molt

Feathers become dead tissue once they are fully developed. They are exposed to wear and tear and must be replaced periodically to sustain the capability of flight, insulation, courtship display, camouflage, etc. Most birds replace their plumage once a year, generally after the breeding season and, in migratory species, prior to migration. Pekin robins have one seasonal molt (prebasic molt) per year in late summer to late fall. The onset is governed by day length, and in breeding birds by the completion of raising their last clutch of young. For individual birds the timing varies accordingly.

Replacement of the entire plumage takes approximately ten to twelve weeks. Birds in poor health or body condition, or with inadequate nutrition, take longer to complete the process. During the molt the birds need extra protein and minerals to re-grow their plumage. They are prone to stress during this time due to reduced insulation, higher energy output for flight, and demands on their metabolism. Good nutrition must be maintained until the molt is completed.

Figure 2.8.1 Pekin robins often drop all tail feathers at once during the molt.

The pigmentation of the feathers is "built in" while they grow. In simple terms, reds yellows, brown, and black colors are produced by pigments or biochromes; blue is created by light refracting due to specific structures of the feather in combination with underlying pigments; and greens are a combination of both forms of color effects. Once feathers are finished growing, pigments can no longer be added to their make-up.

Pekin robins usually shed all their tail feathers at once, but they replace their flight feathers (primaries and secondaries) in succession in a bilateral pattern to maintain balance during flight. Wing or tail feathers that are broken off during shipping or other circumstances may render a bird handicapped until they are replaced at the time of the seasonal molt. The broken-off shafts can be pulled out to stimulate immediate replacement. Individual feathers can be re-grown in four weeks or less.

Young birds replace their mostly gray, juvenile plumage with colorful adult plumage at the age of six to eight weeks, also in the fall. The extended upper tail coverts show their white edges more distinctly in the males, and may provide some clues about the gender of the bird (2.1).

Some bird species have a partial molt to produce special feathers for courtship (for example, egrets). This is not the case for pekin robins; however, under certain circumstances we may see another type of molt, the fright molt. As the name implies, it occurs when the birds are severely frightened, causing them to relax the tissue holding their feathers, releasing them all at once. I have experienced the fright molt with a pekin robin I accidentally caught by the tail, and with a European robin that I discovered to be missing its tail, although I could not establish the cause in that case.

Fig. 3.0-1 Pekin robin aviaries fit well into an East Asian-themed garden.

3

Housing

3.1 The Enclosure

Note: The broad definition of a birdcage embraces a wide range of enclosures; however, in the context of this book these definitions are applied:

- A **birdcage** or **cage** is a traditional, small, barred or screened cage that can be carried by hand.

- An **aviary** is a walk-in enclosure of a floor space of no less 20 square feet and 7 feet high (approximately 2 square meters and 2.10 meters high).

- A **breeding aviary** is a planted enclosure of no less 50 square feet and 7 feet high (approximately 5 square meters and 2.10 meters high).

The real joy of caring for pekin robins and other softbill birds is to experience their fascinating natural behavior, and this is impossible to observe if they are kept as solitary birds in barren cages.

Some literature suggests that a cage for a pair of pekin robins should be 24 x 32 x 24 inches high (60 x 80 x 60 cm). While a captive-raised pair may cope for some time in such a limited space, they will not breed, and will have a poor quality of life. A pekin robin may relent over time and apparently accept confinement in such a small space, but it remains coiled to explode if it is overcome with sudden fear. In an aviary, however, it can dive for cover and remain motionless for a period of time until it feels safe again. In a cage this can lead to injury and severe stress.

It is well known that pekin robins need considerably more exercise than a typical cage can offer. These birds need an environment large and varied enough to provide mental stimulation, a sense of security, and ample opportunities to exercise.

Figure 3.1-1 Never hold pekin robins in a budgie cage.

Figure 3.1-2 Ornate cages are often used in Asia to hold song birds.

There are many bird keepers of the old school who still maintain some softbills in small cages, and whose birds experience longevity and will reproduce in such environments; this is not the case with pekin robins and their close relatives.

I also recognize that as part of cultural heritage, particularly in Asia, many softbill birds are highly prized as songbirds and kept in very small cages. The birds are no doubt revered, which is reflected in the creation of beautiful, ornate cages for them. However, I am not knowledgeable about the custom and technique of maintaining birds for that purpose, and am in no position to judge it from that perspective.

The practice of letting the birds out of their cage into a room to give them longer distances to fly does not compensate for small, inadequate housing. Free flight in a home is dangerous for the birds, and full of hazards that can cause injury or death.

The best way to keep pekin robins happy and safe is to create a suitable habitat in the first place. Condemning a pair of pekin robins to a small cage is like keeping a dog on a chain — full breeding potential will not be achieved.

One exception to this rule is for new arrivals. Initially they are better housed in a confined space in order for the bird keeper to monitor food acceptance, and for the birds to regain their overall balance before they are released into a large, unfamiliar aviary (chapter 5.9). Many birds have succumbed when immediately released into a large aviary where they are unable to find the food dishes, fear attack by other birds, or they simply fade away unnoticed.

Hospitalization is done in restricting cages as well (chapter 11.1).

This book focuses on breeding pekin robins, which requires suitably sized aviaries. Times have changed in aviculture and the traditional practice of caging of wild, rare birds is being replaced with conservation breeding. We must make our best effort to convince bird fanciers of the importance of conservation breeding for this and other endangered species (chapter 10).

Only a few years ago there was no real need to breed this species; when pekin robins were caught in the wild and exported from their native homeland(s) as companion birds. However, it does not take a crystal ball to predict that more and more countries will protect their indigenous, wild fauna from commercial trade by duplicating protective legislation already enacted in North America and Europe. Concern over disease transmission alone has shut down international import and export of wild birds for extended periods and more of this is to be expected.

The availability of a wide spectrum of species will diminish, especially for softbills. Fortunately many hookbill and hardbill species, such as Psittacines and several species of exotic finches, are now domestically raised in impressive numbers to ensure that there will always be birds to enjoy in and around our homes, and in public aviaries. Softbills do not, at this time, enjoy that prospect with any certainty.

The size of an enclosure is only part of the equation. Other elements include the perimeter walls, the interior components such as sight barriers, properly placed perches, a bird bath, climatic control, the location with respect to noise, air quality, sunlight, human and predator activity nearby, interaction with the keeper, hygiene, food presentation, and a host of other factors forming a chain that can break at the weakest link, regardless of how well all other needs are met.

The best guide is nature itself and knowledge gleaned from related behavior study. To replicate nature and natural habitats puts us on the right track to ecologically based aviculture, but it must be done with common sense and logical orchestration.

Figure 3.1-3 Large numbers of finches are domestically bred.

3.2 Small Support Enclosures

A small cage is useful for quarantine, acclimation, holding birds for shipment, health care, and other similar situations. For pekin robins and other species that live at shrub level or the understory of a forest environment, a box cage (figure 3.2.1-1) is the best choice. The birds like the closed-in space, which simulates the plant cover to which they retreat when they feel threatened. However, the front of the cage needs the right kind of screening. As mentioned before, if a barred cage must be used, the bars should be vertical not horizontal. Vertical bars seem to cause less injury to a bird trying to fight its way out.

Case report

A pekin robin was once delivered to me in an open-sided, shoe-box-sized, wire mesh cage made of ¾ inch (19 mm) hexagon chicken wire. The poor bird inside had such extensive scarring around his face, that it had earned him the name "Quasimodo". The unsuitable mesh wire caused unnecessary, renewed injury and severe bleeding. The carrying cage was only fit for a hamster. A used cardboard cricket shipping box, or even a sturdy paper bag with a few small holes would have been a much better choice for a three-to-four hour journey.

Figure 3.2-1 Wire mesh of the wrong dimension causes serious and permanent injuries.

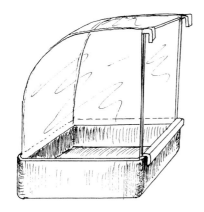

Figure 3.2.1-1 Typical birdbath to attach to the outside of a cage.

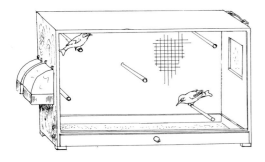

Figure 3.2.1-2 Box cages are best to hold nervous softbill birds.

Figure 3.2.1-3 Matching holding cages make easy cleaning. Two trap/holding cases are used in combination with matching slides.

A covering of fine plastic or metal screen on cage fronts can be installed quickly with the use of a hot glue gun. In addition to protecting the bird from injury, fine screening also provides comfort to the bird by allowing it to be more hidden from direct view.

One-half inch square (12 x 12 mm) welded wire mesh, often called hardware cloth, is suitable for aviary construction, but for small cages this size mesh should be lined or replaced with a finer mesh to prevent the bill and forehead (Figure 1.0-2, topography of a pekin robin) from being thrust through the openings. Metal fly screen is very suitable for covering the sides and the top, particularly if the cage doubles as a trap cage as it allows the birds to see the retained bait inside.

3.2.1 Portable holding cage

Portable cages are 12 x 24 x 18 inches high (30 x 60 x 45 cm) and are carried by hand. They have various uses, for example, for transporting birds short distances, for observation, as shift boxes between inside and outside enclosures (3.3.6) to trap or hold birds temporarily, or, with modification, to restrain birds (chapter 4.1 and 4.2).

A box cage should have a slide-out floor-tray for easy cleaning and, preferably, ways of changing food and water without reaching into the cage to collect dirty dishes for re-feeding. Perches are placed to allow the bird to sit up high and far enough away from the end walls so as not to wear out its tail feathers, with additional lower perches placed to descend to the food dishes. If the bird is kept in this cage for several days it should have a birdbath. One can use two large, covered bird dishes, of the type that are hung on the side of the cage and are sold as birdbaths, and dedicate one for food and one for water.

The birds will bathe in this dish, as well as drink from it, which necessitates regular water changes. It is best to fix these dishes to the solid end-walls of the box cage so that they can be removed without an arm and hand entering the cage and frightening the birds. Few things are more upsetting to them than something or someone reaching into their already limited space.

Here is a valuable tip for nervous softbills: construct two identical box cages with slide openings that line up to each other so that the cages can be joined together as a shift cage.

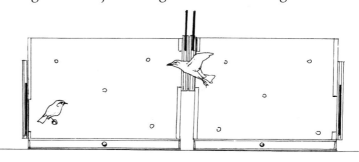

Steps to clean the occupied cage are as follows:
- Place both cages end-to-end tightly together with the slides facing each other.
- Open both the slides.
- Cover the cage holding the bird with a dark cloth (the bird will move or can be coaxed to the bright, uncovered cage).
- Drop the slide on the cage the bird entered.
- Remove the cage that is to be cleaned a distance away to keep the bird undisturbed while cleaning it.

An additional slide track made of two grooved strips of wood, fixed to the sides of the slide opening, allows a small net to be slipped in place, ready to catch the birds rushing to the light, after you darken the cage with a towel. The bag of the net should be long enough to entangle the bird as it enters to allow for time to drop the slide so the bird will not quickly come back out. The captured bird can then be taken into your hand for treatment or transported in the net to another aviary, shipping box (chapter 4.2), or another location easily and safely.

If you try to catch a lightning fast pekin robin in a cage by reaching through the door you may have it come right at you, pass your hand, and fly off before you realize what has happened. If you are used to catching finches this way, you will be amazed how crafty pekin robins are.

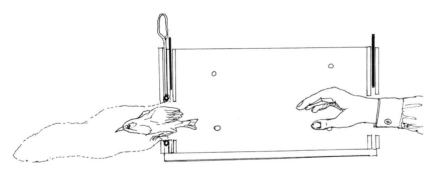

Figure 3.2.1-4 A sock-shaped net is ideal for catching fast birds.

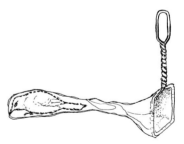

Figure 3.2.1-5 Folding the net blocks the way back out.

3.2.2 Hospital cage
A hospital cage is used for birds that require special treatment, such as high ambient temperatures to regain and sustain body heat, daily medication or hand feedings, recovery from impact trauma after hitting an object, to heal fractures, and other needs. It can also be used for hand feeding chicks when they approach fledging age, until they feed on their own and need more space to develop (chapter 7.5).

The ideal hospital cage should have a slide out floor tray and large, cabinet-style plexiglass front sliding doors to make it easier to change perches, allow placement of food directly in front of the bird, and facilitate thorough cleaning. A triple track can accommodate three sliding doors that will allow division of the cage front into three sections, providing easy access to any part of the cage.

The unit should have a solid back wall, and a ceiling with a vent opening. A practical size for smaller softbills is 12 x 24 x 18 inches high (30 x 60 x 45 cm) as suggested for the portable holding cage noted in 3.2.1 above.

There should be metal screening on one side where an infrared heat lamp is positioned. The heat lamp should be clamped to an independent base to allow it to be moved to different positions in order to establish the correct temperature inside the cage. A thermometer needs to be mounted on the inside wall to monitor cage temperatures. A small fluorescent light source is placed above the ceiling vent to provide "daylight".

A simple way to set up perches is to screw a piece of wood, with mounting holes drilled into it, onto the back wall. The perches can be easily removed and pushed into the holes in any desired location. The holes should be offset to avoid the accumulation of droppings. This is handy when you have an injured bird and like to restrict its movement by removing the higher perches. The perches should not be so long as to touch the front plexiglass doors.

The hospital cage is best painted with a good, washable paint. Paper towels work well as floor covering because they are quickly and easily changed, and show and absorb the patient's discharges well. If you wish to collect a stool sample, wax paper can be placed under the roosting perch. It will prevent rapid dehydration of the fecal sample (chapter 11.2.2).

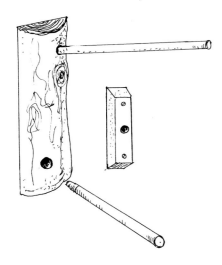

Figure 3.2.2-1 Mounting blocks for push-in type perches.

Figure 3.2.2-2 A practical hospital cage with its many features in front view, side view and an insert showing the triple track of the front sliding doors.

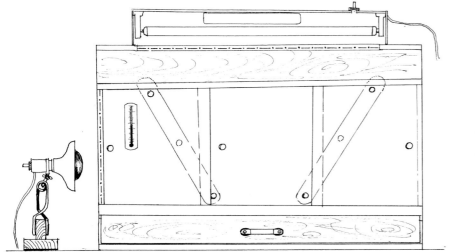

front view

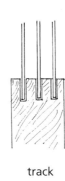

track

side view

3.2.3 Shipping box

The International Air Transport Association (IATA) regulates the crating of animals for shipment by air. You can request specific information for the type of animal you plan to ship from the air carrier. The carrier will provide you with construction drawings, dimensions, and other stipulations for the crating of live birds. The design ensures that the box will be carried right side up, prevents vents being accidentally blocked off by other cargo, and helps to avert undue stress for the birds.

An important modification to make for shipping softbills, which is not mentioned in the drawings of the air carrier, is to line the box front and other openings with fine mesh to prevent facial injuries, as explained above (3.2). If you reuse a shipping box with coarse screen make sure to line it with fly screen on the inside to prevent facial injuries.

The box size depends on how many birds are to be shipped. I suggest a single bird or a pair of pekin robins should have a space measuring 8 wide x 10 deep x 6 inches high (20 x 25 x 15 cm). For additional birds, the width should be increased by about 2 inches (5 cm) per bird.

The box should be made leak proof by placing a bead of hot glue along the bottom seams.

Two perches are essential to allow birds to hop from perch to perch, and thus stay off the floor that is often wet from spoiled water and the birds' discharges. Birds shipped in a box with two perches arrive significantly cleaner and with fewer damaged feathers than one with only one perch. A perching bird air-shipping container with no perches at all is completely unacceptable.

Multi-compartment shipping boxes often have a common lid. This makes it difficult to remove a bird or birds from one of the center compartments. Therefore, a narrow sliding roof panel, instead of a side door, is highly recommended for the safe transfer in and out of a selected compartment. By using a shipping box with a sliding roof panel, in combination with a portable holding cage or other type of box equipped with a sliding floor panel, birds can be moved out of individual compartments without being handled.

We can take advantage of a bird's desire to move to more open spaces, particularly if they have more light. To accomplish this:

- Place a small holding cage with a sliding floorboard on top of the shipping box.
- Open both the holding cage floorboard and the shipping box lid (or top slide) to create an opening large enough to allow the birds of one compartment to fly up into the holding cage.

Figure 3.2.3-1 A small holding box with a sliding floorboard works well to selectively remove birds from a compartmentalized shipping box.

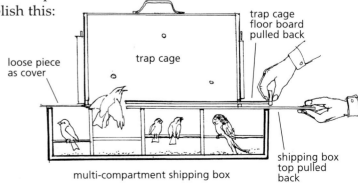

- Once the bird(s) move up into the holding cage, slide the floorboard shut.
- Close the shipping box roof panel to keep remaining birds inside and the transfer is completed.

Provisions must be made to allow for emergency feeding and watering. Food and, more importantly, water dishes must be installed inside the box. To accomplish this, holes large enough to permit serving the dishes inside the box but not large enough to allow escape, are cut into the front screen above each of the water and food dishes. It is a good idea to place a piece of tape over the holes so birds do not force their beaks through them trying to work their way out. The dishes need to be glued in place.

For short trips of less than eight hours, water can be omitted as long as thick slices of oranges and apples are placed in the box. For uncertain situations, the water dish should be filled and a slice of apple or orange floated on the top to keep the water from splashing out. A piece of sponge can also be used for longer trips to keep the water fresh and prevent spillage.

There should always be slices of fruit available for the bird in case the box is accidentally put on its side, or in the unlikely event that it is turned upside down (heaven forbid!) during shipment, so that the birds have access to liquids. This is also of great importance in case the shipment is delayed.

Chick starter crumbles mixed with bird seeds can be used for bedding. It absorbs water and can also act as emergency food.

It is good practice to tape a two-to-three day portion of non-perishable food to the box to enable the consignee to gradually put the birds on the diet he/she will be offering. This practice will also allow air transport personnel to feed the birds in case of an emergency caused by unexpected delays. General feeding instructions, diet recipes, and ID information can be emailed ahead of time to the consignee and/or taped to the side of the box.

I have painted my boxes fluorescent orange for visibility. So far none have been lost or overlooked in transit.

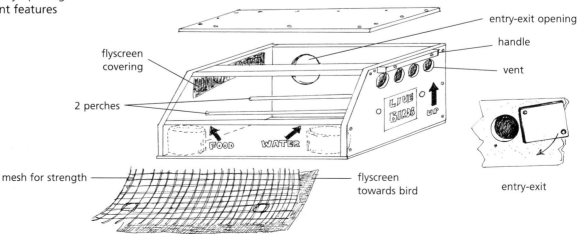

Figure 3.2.3-2 Fine screen, two perches and an exit/entry opening are especially important features for softbill birds.

3.3 The Breeding Aviary

Success in nesting and rearing chicks increases with the size and habitat quality of an aviary. The term "aviary" has been applied to a variety of enclosures, from something big enough to allow a bird to flap its wings while hopping from perch to perch, to magnificent walk-through aviaries replicating expansive, natural habitats.

The term "flight" is often used to describe and market birdcages that are larger than handheld cages, but often less than three feet (91 cm) in length. While it may be a comforting thought for a bird keeper to have acquired a "flight", it is not necessarily an enclosure that actually allows real flight for the birds inside.

In essence, for the best results we must endeavor to create a functional habitat that resembles natural features found in the wild. It will be repeatedly stated in this book that the best breeding aviary is one that is sufficiently large, well planted, and dedicated to only one pair of softbill birds. We should never lose sight of these three parameters.

The creation of a "biotope aviary" is an art form in itself, linked to landscape architecture. A beautifully landscaped aviary will significantly enhance the home and property of its owner. The propagation of birds in such an environment is entertaining, educational, and spiritually satisfying. A pekin robin aviary can fit beautifully into many garden settings, especially one with an East Asian theme (figure 3.0-1).

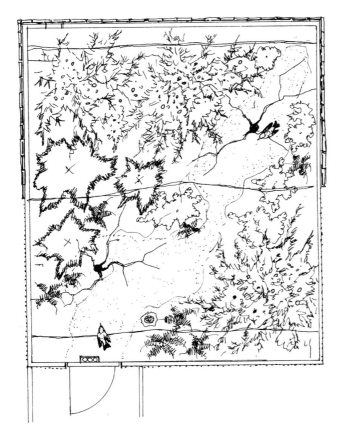

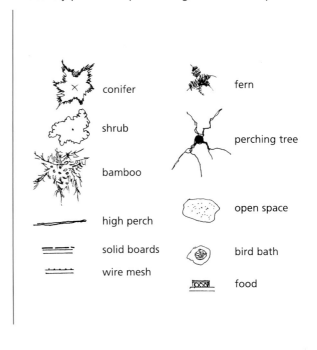

Figure 3.3-1 Prototype of a breeding aviary; note dense planting, solid back wall portion, feeding tray served from the safety porch and positioning of horizontal perches.

3.3.1 Size and configuration

As pointed out in the introduction of this chapter, an "aviary" is defined as a bird enclosure at least big enough for the keeper to step into, stand up in, and turn around in, no less than 4 x 5 x 7 feet high (1.20 x 1.50 x 2.10 m) or similar floor space configuration. While it could be slightly smaller in its floor dimensions it should be seven feet high to give the pekin robins enough height to feel secure when flying to the high perch. Few pekin robins will breed in this size enclosure, but it is a start, and with supportive development (dense plant cover) of the interior for nest sites a pair may settle down to nest. To maintain or display two to six pekin robins, the above size aviary is adequate. For a standard breeding aviary I recommend a floor space of 6 x 9 feet (1.83 x 2.74 m) (chapter 6).

For an outdoor habitat aviary, which requires more landscape work, I try not to build anything smaller than 6 x 9 x 7 feet high (1.83 x 2.74 x 2.13 m), because it is just too difficult to move around among large plants with garden tools within a smaller space.

On numerous occasions I have found that pairs displaying nesting behavior in smaller enclosures (but did not complete building a nest) did lay eggs and care for hatched chicks after they were moved into larger aviaries. Conversely, a pair that constructed a nest and was then moved from a large enclosure to a smaller one did not resume nesting until being moved again to a larger space.

On the other hand, "the bigger, the better" is not necessarily true for breeding pekin robins. While a larger aviary can be made into a more impressive habitat, with the added benefit of establishing a walking path for easy maintenance and the luxury of being among the birds, it can pose problems for monitoring them on a regular basis, or for capturing them. Pekin robins will not tolerate a second pair in their territory, either their kin or another closely related species. While they may not display obvious signs of territorial aggression with less related species, breeding success has been inconsistent in community exhibits (chapter 10.1).

Ethologists speak of a "critical flight distance". This is the distance at which the animal will take flight or flee from an approaching or perceived enemy. Another term is "fight distance," that point at which a cornered animal decides to reverse direction and attack. This does not directly apply to small songbirds, but don't be surprised if the parent pekin robins attack you when you try to catch their loudly complaining fledged chicks.

Theoretically a circular aviary that has screened sides all around would need a radius of no less than the flight distance of the particular birds inside to have a comfort spot in the very center; "particular" because each bird has its own distance based on level of trust and experience. Most pekin robins relax at a distance of three to four feet (0.9 to 1.2 m) from the aviary front, unless a net

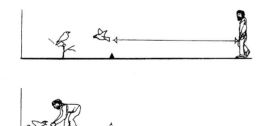

Figure 3.3.1-1 Flight and fight distance.

wielding "enemy" has recently chased them around. So, for example, a bird with a four-foot flight distance would need an eight-foot diameter, all mesh-sided aviary, and it would still have only one comfort spot in the very centre. A bay or horseshoe design saves space and stress. By enclosing three sides, at least partially, we avoid the need for a large aviary to give the birds a stress-free comfort zone.

From a functional point of view, breeding aviaries only need to be "open" at the front. To allow more sunlight to reach into the aviary environment one-third to one-half of one side may have mesh wire as well.

Dense vegetation and a high perch at the far end from the viewing front are spots of retreat. For smaller breeding aviaries it is best to employ the boxed-in effect. It simulates dense bush and extends the birds' comfort zone. Small side-by-side aviaries need solid, opaque partitions between each pair of pekin robins so they can settle down to breed (chapter 6.5).

A portable, round, or multi-sided aviary with screen on all sides can be placed into a corner of a room and the sides can be blocked with potted plants to simulate the boxed-in effect. Correspondingly, as the floor space of a planted aviary increases, the need for closing in the sides decreases.

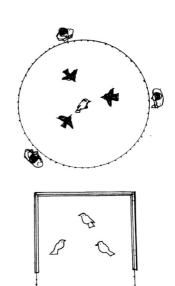

Figure 3.3.1-2 Proportionate comfort zone space of aviary types with equal foot print.

For an aviary, ½ x ½ inch (12 x 12 mm) screen works well as a barrier to keep the birds in and at the same time, adult mice out. To stop juvenile mice from entering, a smaller mesh must be chosen. To improve the transparency and to arrest corrosion it is practical to paint the galvanized wire mesh with flat dark paint by using a paint roller. For an indoor aviary a 3/8 x 3/8 inch (9 x 9 mm) mesh is a better choice to control visits by mice. Painted, metal fly screen is another option to contain crickets and other live food inside an indoor aviary. Otherwise insects should be killed just prior to offering them to the birds. This is labor intensive and demands more frequent feeding than offering insects in the self-serve feeder (chapter 6.10).

3.3.2 The floor

The floor of an indoor aviary can be made of various materials, keeping in mind the introduction of live plants. Plants can be potted for a smaller space with solid floors, or placed in planters for larger units. With adequate drainage, a permanent, indoor aviary could have a wash-down floor.

However, washable substrates such as concrete, ceramic tiles, and paving blocks require a frequent hose-down to remove the waste. This can present a problem, especially in small aviaries, as it upsets the birds too much, particularly during breeding season. To reduce disturbances we can add bedding including sand, cat lit-

Figure 3.3.1-3 A solid partition is a must between breeding territories.

ter, peat moss, and wood shavings, thereby avoiding daily cleaning. Bedding must be managed to keep it clean, dry, and dust-free.

For an outdoor aviary natural soil substrate is the simplest material, since the plants can be cultivated in it. For an aviary with a dense bird population, however, soil is not ideal. Such an aviary needs a lot of substrate replacement in order to keep it sanitary. Sand is a better option in that case; it offers better drainage and reduces compaction of waste more than soil does.

The use of sand or natural substrate is quite workable due to the low number of breeding birds per aviary space. The accumulation rate of waste is low enough to allow maintenance work to be delayed during critical nesting periods.

3.3.3 The walls

It makes sense to standardize the wall dimensions for a series of aviaries so that they are modular. This way they are interchangeable and quickly assembled with screws, not nails. The walls must be set on a foundation to keep the wood away from moist soil that would rot the framing in time. Stone, concrete block, or concrete poured in place are the preferred materials. I have used heavy western red cedar *Thuja plicata* planks 2 x 12 inches (5 cm x 30 cm) wide or wider, treated under the soil line with wood preserver and painted with oil stain above it. This permits fast construction and moving or enlarging the structure if need be.

The sidewalls can be of wood framing with mesh wire fastened onto it. I generally place the framing on the inside of the walls. Inside frames create flat walls on the outside, which thwart the neighbor's cat and raccoons with aspirations of scaling the sides of the aviary. In areas where rodents, and rats in particular, are a problem the mesh wire should be nailed to the outside to curtail their attempt to chew thorough wood frames and boards.

A "pet guard" may be desirable. This is a short solid section along the screened front and sidewalls designed to create a sight barrier for cats and dogs strolling by. It reduces stress for the birds and their keeper.

It could be argued that framing on the inside is prone to collecting debris. While this may be so, it is negligible, because of the few birds held in the relatively large space. Pekin robins do not like to perch on the wire if given a suitable perch. They do like to sit on the ledges of a sun-drenched wall to sun bathe.

Predators that dig under the foundation are a major concern. Rats are by far the most troublesome. A reliable guard must be buried around the perimeter. Depending on soil condition it may have to be as deep as two feet (60 cm) or more and create an outward sloping barrier. Most people use heavy, galvanized wire mesh. Rats, if they gain access, will make nesting impossible and even kill the adult birds at night.

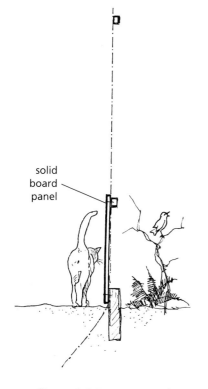

Figure 3.3.3-1 Pet guard.

A strong wire mesh floor makes planting shrubs and trees into the ground difficult. A sufficient layer of soil must be on top of the wire to anchor the roots at planting time. Potted plants are the way out of that dilemma, although create another challenge in making the setting look natural.

You may ask why potted plants are so important. I never had a pair of pekin robins nest in anything but live plants. Of nearly one hundred nests my pekin robins and mesias have constructed only one was built in a bundle of dry bamboo straw, but it was next to a live conifer. If the birds were reluctant to build a nest, the introduction of additional live plants, mainly bamboo and conifers, inspired them to nest.

I have not tested whether the birds will accept artificial plants, and other dead substitutes for live vegetation, but must assume that at least some pairs will. Knowing of a good number of people who have tried unsuccessfully to breed pekin robins without planted aviaries gives no encouragement to experiment with this; besides, the birds look so splendid among live plants where they find more things to investigate.

Panes of glass may be part of a wall section or the roof. Birds must be conditioned to this form of barrier to avoid serious injuries by flying into them. In the past I have used wet potter's clay applied to a rag and wiped over the glass to make it opaque or "blind" it. The clay can be gradually removed, but must be re-applied when new birds are introduced and chicks begin to fledge. A bar of soap can also be used but birds may come into contact with it, which might interfere with digestion and the protective oil of their plumage.

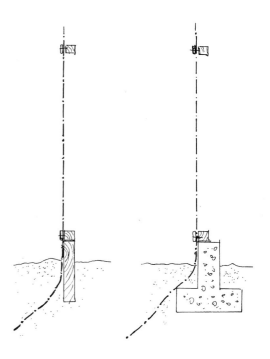

Figure 3.3.3-2 Aviary wall cross-section with dig-out curtain (rodent guard).

A word of caution:
Startled birds may still fly into a bright opaque window, in particular newly fledged birds and especially silver-eared mesias, which seem to explode and take off in "headless" flight towards a bright window. A pestering hawk or a stranger hastily entering the aviary can trigger such events.

I recently lost two healthy mesia chicks despite "blinded" windows. I solved this problem by placing mountain hemlock *Tsuga mertensiana* branches in front of the glass. The filigree of fine branches creates a buffering, net-like barrier to which the birds react visually and, if that fails, by contact. The branches are permanently installed, look natural, and allow viewing of the birds, including with binoculars, especially after the needles have dropped off. The branches break up the image of the keeper on the other side of the glass to serve as sight barrier for easier observation and comfort of the birds.

An alternative to live branches is nylon bird netting, sold to protect berry crops, which is stretched firmly and tightly onto a

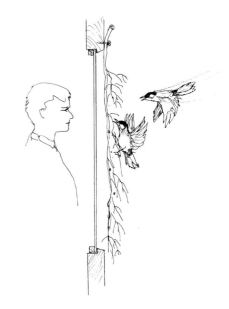

Figure 3.3.3-3 Conifer branches prevent glass impact trauma.

frame. I use this for a patio door leading into a walk-through aviary in place of the usual sliding, fly-screen door. It allows perfect viewing and has prevented any impact trauma.

3.3.4 The roof

The roof should have a rainproof cover over approximately half of it with the option to cover the remainder, which may become necessary if the birds decide to nest in the unprotected part of the aviary. There should always be a cover at each end of the aviary, since birds fly from end to end during normal activity. The end perches are critical stopping points and if they are exposed to pouring rain, bird plumage can become drenched.

Case report

I once released six newly arrived coal tits, Parus ater, to an outside enclosure (which was actually a protective cage for a bed of raspberries to safeguard them from plundering birds) to condition them to more exercise and outdoor exposure.

A sudden rainstorm soaked the birds to the point where I had to pick them off the ground, unable to fly, and bring them inside to restore normal body heat and functions. The birds' plumage had become damaged during air shipment (in a cramped box without perches!). As a result they were incapacitated by the storm and, still unfamiliar with their surroundings, they did not know how to return to their indoor shelter. Had I not been right there, the birds would have suffered severe body heat loss and could have perished. The birds weathered any rain thereafter once their plumage was reconditioned, and both ends of the enclosure were covered.

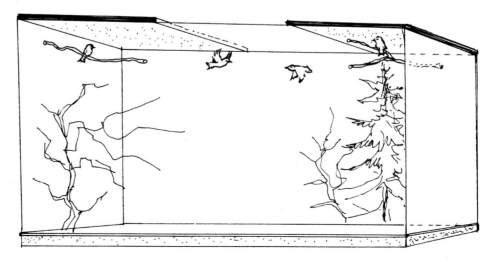

Figure 3.3.4-1 Covered ends of a flight allows birds to perch out of the rain.

The simplest approach is to cover the entire roof with translucent material that will allow light to filter into the aviary and promote plant growth. It is frustrating to climb up on the roof to place rain covers over the nest sites, only to put the birds off and then have to move the covers to the next chosen sites. Rain pelting down on the nestling — albeit covered by a brave, drenched parent holding off on feeding — and turning the nest into a soggy mess may be nature's way, but is not a good way to test the survival strengths of your birds. Besides, nests are lost this way in nature as well.

A covered roof eliminates wild bird droppings, attacks by predators from the air, cats from reaching through to the high perches and, if semi-translucent, prevents the extreme heat of direct sunrays from reaching the nest. Watering the plants can be conveniently arranged with irrigation drip systems (3.3.15). It ends the worry about soaking the nest, or chicks that just fledged.

I have found corrugated fiberglass panels to be problematic as roofing material for a breeding aviary. A neighbor who used this type of roof had a problem with stray cats climbing up on the roof. The considerable noise made when they walked on it created havoc at night among the pekin robins, causing them to bolt and be kept on the move, resulting in injury and even death. In one case the brooding hen spilled the eggs and the male was found dead in the morning with a broken neck.

Secondhand, tempered-glass panes — removed from patio doorframes — work well as rain covers. Patio glass doors have sealed double panes of tempered glass. For this purpose the seal is removed to obtain single panes of glass. They are heavy and stay in place in heavy winds, and can be slid aside to allow rain in during the off-breeding season. UV light, which is blocked by glass, should still be able to reach the aviary interior through wire mesh side panels. Glass panels should not enclose the entire aviary.

The clear glass panes should be painted with a semitransparent stain to cut heat build-up behind them. Cats apparently do not like to walk on glass, but if they do climb onto it, they do not create as much disturbance. Stray cats, however, must be removed (chapter 4.3.3).

To further avoid build up of heat under the glass roof, we can prop up the glass to create airflow under it. This is assuming that the roof has a wire mesh ceiling below. (In fact, I insist that it does because even a tempered glass pane can break.) A piece of plywood can also be placed in the right location to provide additional shade for the nest during the hottest sunshine hours.

In 2006, the alarming cases of avian influenza in various parts of the world spawned new concerns for aviculturalists. Bio-security measures call for protection against potentially contaminated excrement of wild birds that might find their way into aviaries.

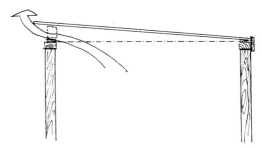

Figure 3.3.4-2 Raised glass cover to improve venting in hot weather.

This measure alone is a convincing argument for a complete aviary roof covering and secure run-off away from the structure. Commercial fowl, domestic or pet birds, and even humans are other possible vectors and, as a result, many preventative protocols are under development at this time.

3.3.5 Safety porch

A safety porch, through which the keeper can enter the aviary, is a must with pekin robins. One must never enter the aviary through a door that leads to the "great outdoors" with the birds present. This is playing Russian roulette with bad odds.

And if you do have a safety porch, it is only useful if you close each door behind you — always! I have installed self-closing devices on all doors by connecting a bungee cord or a rubber band cut from a truck inner tube between the door and the frame. It has prevented escapes more than once. Spring-loaded hinges can also be installed.

Indoor/outdoor aviary complexes can solve escape problems by entering via a service corridor, but doors must still be closed.

Figure 3.3.5-1 Feeding from a safety porch "through the wall".

3.3.6 Shift boxes

A shift box is a blessed invention. This is a small cage through which the birds are moved from inside to outside aviaries, or other connected spaces (Lay-out Plans, 3.3.7). The size may be 12 x 24 x 16 inches high (30 x 60 x 40 cm). The shift box has slides at each end that are also used to block the bird's passage when needed to trap it for capture, observation, treatment, or other needs. By making them removable they serve as trap, transport, or transfer cages to load a shipping box, for example, all without ever needing to handle the bird. This form of capture is more creature friendly than chasing the poor bird to exhaustion, finally scooping it up with a net. Stress must be avoided in managing these valuable birds. Succeeding in that is a win-win solution.

The boxes should be standardized to make them inter-changeable with a back-up box for thorough cleaning, repair, and repainting.

A word of caution:
Birds can be accidentally locked away from their food if we lose track of open and closed slide doors. To avoid this, the string-pull of the slide should have a large label, which hangs in clear view as a reminder when the slide is shut. Any unexplained signs of untouched food must be followed up to confirm that the birds have access to the feeding station. Unfortunately, even with all the much-appreciated benefits of shift boxes, I have, with great sorrow, lost birds that way.

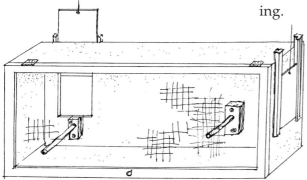

Figure 3.3.6-1 Shift boxes connect inside with outside spaces and double as trap and transfer boxes.

3.3.7 Lay-out plans

An aviary system for several breeding enclosures must have a service corridor on the inside of the building as well as on the outside. It prevents escapes through access doors while carrying out maintenance of the aviaries. Connecting corridors also facilitate moving birds from enclosure to enclosure without the need for capture.

If you should happen to be in the enviable position of starting new construction, rather than having to add onto an existing structure, it would be wise to incorporate options for expansion into the initial building plans. Inevitably, if you breed birds, you will be looking for more space.

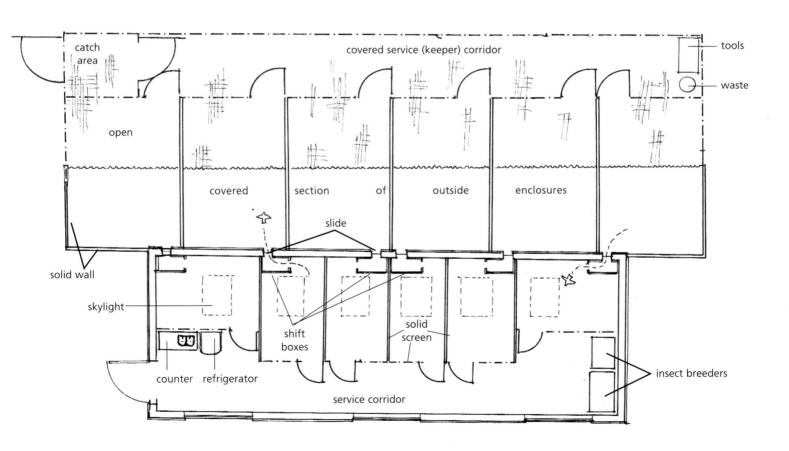

Figure 3.3.7-1 A six unit aviary system.

A common shelter can serve two breeding aviaries nicely. The service area becomes the safety porch with double doors for security. The passage for the birds between the indoor and out door flights takes place through removable shift boxes, as noted above.

**Double Unit
Indoor-Outdoor Aviary**

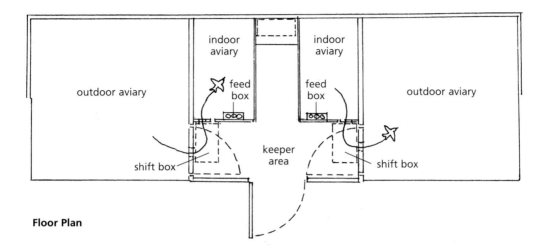

Floor Plan

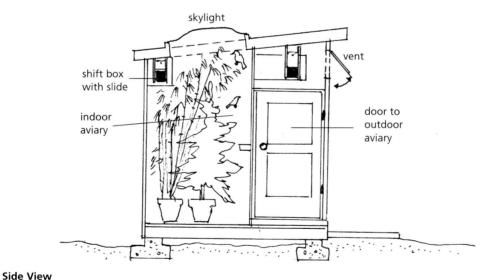

**Side View
of Indoor Shelter**

Figure 3.3.7-2 This double unit uses the service area as a safety porch for all aviary doors. Birds move in and out through shift boxes. A skylight aids plant growth.

Chapter 3 ■ HOUSING

A single breeding aviary can be built in sections to facilitate transport and assembly at the designated site. The space above the safety porch is utilized as a sleeping loft, with a remote controlled slide to shut the birds in if required. A small unit requires good buffer planting to create privacy.

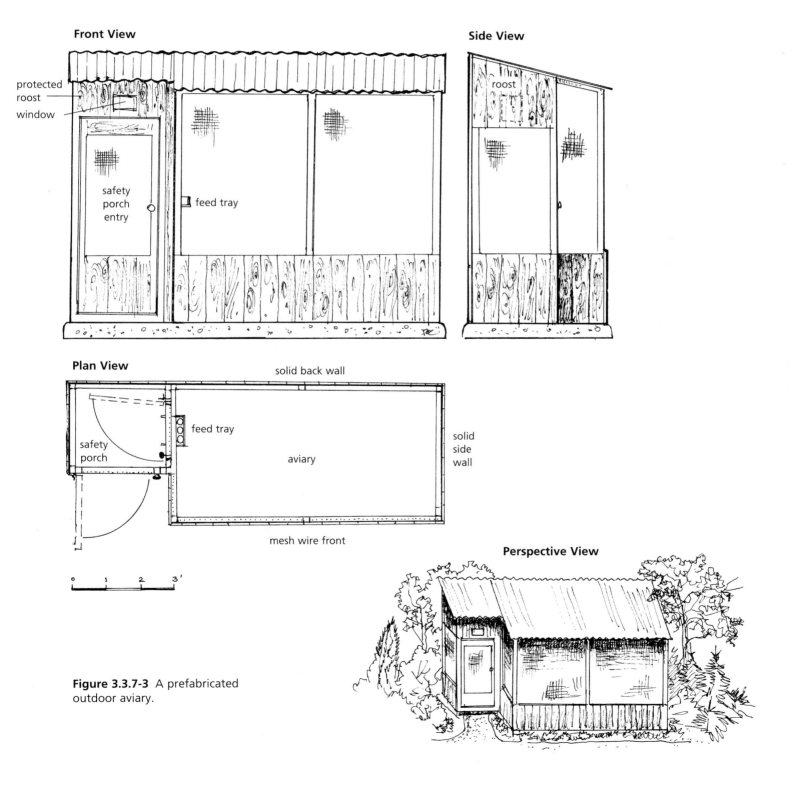

Figure 3.3.7-3 A prefabricated outdoor aviary.

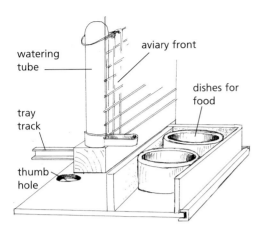

Figure 3.3.8-1 A feeding station with a drinker and a slide-out tray for food dishes.

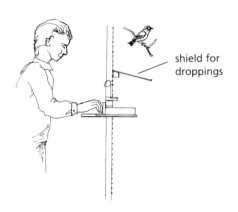

Figure 3.3.8-2 A protective shield prevents contamination of the food station.

Figure 3.3.8-3 A small back-up feeder for dry food is good insurance.

3.3.8 Feeding stations

Feeding stations are worth reviewing since feeding is the most important daily activity. If you have a number of aviaries you will be interested in improving the efficiency of daily routines. Specially designed feeding stations that allow for quick and safe feeding, particularly for a relief person, without the need to enter the enclosures, are highly recommended.

The food is provided in glazed, ceramic dishes, which are set in slide-out trays. A catch can be added to prevent the tray from being pulled out entirely and to hold it in a position that blocks the opening, to keep the birds in. The trays should be standardized. A back up unit should be available to switch a tray for thorough cleaning or repainting. There are double sets of ceramic dishes all of the same size, to allow one set for feeding and one for cleaning. Softbill food spoils during hot weather and becomes a breeding ground for bacteria and fungus spores; hence daily cleaning with hot water is a must.

The correct number of dishes should be filled in the food kitchen and taken to the enclosures. If one is left over, then one aviary has not been served. Softbill birds need to be fed daily. None may be left without food or water, ever! Overfeeding or feeding in self-feeders, as can be done with seedeaters, is not an option for pekin robins, which are provided with perishable food.

The right feeder design and an established routine make caring for the birds easy. Where necessary a shield should be mounted above the food station to keep bird droppings out. Such a shield can also be used above the water bath. A piece of clear plastic works well.

In most environments feeding stations attract mice. To catch them, the food tray should be pulled out at an angle, only allowing the mice to pass through and not the birds. The dishes should be removed at night so that the mice, arriving at their familiar feeding place, can then be easily trapped (chapter 4.3.1).

A word of caution:

If your feeding routine is prone to interruptions it is possible that a tray with fresh food does not get pushed back into position for the birds to have access to it. A final check on all feeding stations is recommended. I lost a healthy silver-eared mesia in less than twenty-two hours on a near-freezing day, when energy demands are particularly high, due to such an accident.

Another idea is to place dry, non-perishable food pellets in a back-up feeder that can serve as an emergency ration in a place that is independent of the feeding station. A dish with crushed eggshell can be positioned at the back-up feeder. It must have a cover to protect the food from rain and debris.

3.3.9 Birdbaths

Birdbaths can be positioned on the wall and fed by a network of irrigation drip lines, with overflow drain lines to the outside of the building or the floor drain. In outdoor aviaries the lines can be left on with a small drip to water the plants. In some cases I mount the supply tube to a stake that can be moved, together with a ceramic water bath, to distribute the watering function in the planted aviary.

Glazed ceramic dishes, sold as flowerpot saucers, can be drilled with a glass drill to connect an overflow nipple for the drain tube, which is fixed in place with epoxy glue. A series of water dishes can be hooked up to activate the water flow centrally for periods of time to replenish the water.

As a form of insurance, drinking water is provided in drinking tubes next to the food dishes, since the happily bathing pekin robins may splash out all of the bathing water (chapter 8.9).

Figure 3.3.9-1 Small water tubes charge and discharge the birdbath.

3.3.10 Perches

We see pekin robins readily alighting on a vertical or slanted bamboo stalk, or watch them choose an obscure branch in dense bush, and we appreciate their agility and adaptation to the bush type of environment. However, a horizontal perch is the preferred landing site for most birds and is, in fact, irresistible to Pekin robins.

The placement of perches is more of a science than one might expect. For example, I found that a series of horizontal sticks placed in a certain way lead the birds along the service corridor to another part of a building. Perches lure and transfer the birds through unfamiliar spaces with little hesitation on their part. Likewise, if birds are not willing to come out of their aviary door into the service area, placing a stick in front of the open door works like a magnet to entice the birds to leave their familiar, safe enclosure.

Within the aviary, perches should be placed at the end walls as far away and as high as possible without causing the birds to rub their tails when they turn around on the perch, or to be so visible that predators can see their outline through a transparent roof.

The birds exercise by flying from one high perch to the other, and like to do this over the longest distance. If the perches are not close to the end walls they fly past them and land on the wire. This not only roughs up their elegantly outward curving tail feathers, but it attracts watching cats and birds of prey. The predators may not be able to get hold of the bird, but they can certainly cause serious injury.

If a roof is not covered with solid material and only has a wire mesh barrier, then the perches must be low enough to keep the birds beyond a reaching cat's paw or bird of prey's talons. Pekin robins have the habit of roosting very tightly together at night and

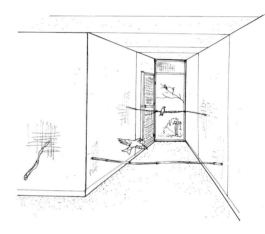

Figure 3.3.10-1 Temporary, conveniently placed perches work like magic for moving birds to a desired location.

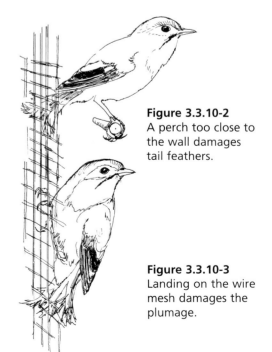

Figure 3.3.10-2 A perch too close to the wall damages tail feathers.

Figure 3.3.10-3 Landing on the wire mesh damages the plumage.

Figure 3.3.10-4 The placement of perches dictates the activity of birds and aids clean up.

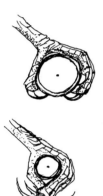

Figure 3.3.10-5 Proper diameter perches help to keep the toenails trim.

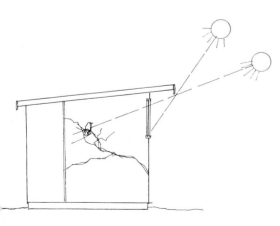

Figure 3.3.11-1 A window high under the eaves allows the winter sun to shine into the building but blocks the hot summer rays.

only a major disturbance makes them bolt. This makes them vulnerable to owls and cats, and also rats as mentioned before. A few strategically placed conifer branches on the outside alleviate the problem.

The two high perches are the most important. Others can be placed to break up the remaining space. The placement of convenient perches dictates the movement pattern of the birds within the aviary. This is helpful in directing the birds to places where the aviary floor can be covered with patches of sand rather than vegetation to make the removal of excrement easier. Placement of perches in shipping boxes requires care as well (chapter 3.2.3).

Natural branches with varying diameter are the best choice to ensure healthy feet and controlled nail growth. Fresh-cut branches are better than hard, dry ones for reducing impact on the ligaments when the birds land. Metal pipes or plastic covered metal rods should not be used.

Several perches should be thick enough so that the pekin robins cannot reach around the perch with their toes. The nails should stay in contact with the perch; otherwise the toenails will not wear down properly. This is of greater concern for birds held in smaller enclosures that are not landscaped. Clearly, a perch is much more than a stick in a cage.

Finally, it is good practice to observe the birds from time to time to establish where they go to roost, to confirm their safety, and to be able to locate them in the dark with a flashlight should it become necessary to make a head-count.

3.3.11 Positioning the aviary

Ideally, the aviary should face southeast to catch the warming morning sun. A direct southern exposure benefits from the dappled shade of a tree to buffer the hottest hours of sunshine.

Strategically placed shading is important in the breeding season. The adults can brood the chicks to keep them warm, but not cool them. An attached shelter should have its windows near the roofline to block summer sun and yet give access to low-angled rays in the winter.

The bird breeder must be aware of temperature fluctuations within the aviary because direct sun behind glass can be deadly for chicks in the nest. A thermometer must be placed and monitored in all aviaries where it is suspected that the temperature might vary drastically.

Wind protection is a must and additional cover must be installed to keep prevailing wet winter storms out. A rain-soaked bird in a drafty environment is much worse off than one in a still, dry, and cold one several degrees below freezing.

The aviary should also be located where there is the least

amount of disturbance, even though birds learn to accept this if it does not affect them directly. Most resident pets that are familiar to the birds are soon ignored if they do not harass the birds.

Planting shrubs, vines, and trees around an aviary extends the territory of the birds psychologically; it attracts additional food insects, reduces predator attacks, buffers temperature fluctuations, and makes the structure more attractive in garden or park setting.

3.3.12 Winter housing

This book reflects my experience in breeding and housing numerous pairs of pekin robins in outdoor aviaries on the West Coast of British Columbia, Canada. While protected indoor rooms are provided where water and food is protected from freezing, the pekin robins and silver-eared mesias often insist on roosting outside. During the winter months the temperatures may drop to several degrees below freezing overnight. In prolonged subfreezing weather the birds can be shifted and locked inside. Adequate food supply is absolutely critical in cold weather (chapter 8).

Figure 3.3.12-1 Good cover helps the birds to choose a night roost.

An ideal facility has an indoor holding space connected to an outdoor aviary to allow birds to stay in their familiar territory year round. In milder, frost-free regions the birds can be left to their own devices in such enclosures in all seasons, as long as there is good wind and rain protection. In colder climates it may be necessary to lock the birds inside a space that is above freezing, particularly during the night.

Pekin robins can be trained to roost inside by chasing them into the inside enclosure for several consecutive nights and forcing them to seek out an indoor roost. Most of the time the birds will change their habit and accept the new inside roost and no longer need to be locked in. Dense vegetation reaching high under the ceiling or roof is a key factor in selecting night roosts. Adding more green branches or other forms of vegetation to the inside will usually entice them to stay in.

The feeding stations should always be located in the inside enclosure to habituate the birds to coming inside on a regular basis, and also to provide better protection from rain, direct sun, and freezing weather to the food itself.

In colder climates an indoor/outdoor system allows birds to be let out on mild weather days. The window of opportunity for successful nesting may be narrower where winters are colder and the birds must be kept indoors until late spring. Breeding behavior will commence with increasing daylight (vernal equinox) even without access to planted outdoor breeding aviaries.

3.3.13 Running water

As discussed earlier pekin robins are attracted by moving water.

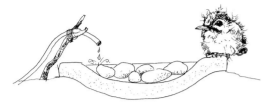

Figure 3.3.13-1 Pebbles in a deep bird bath prevent fledglings from getting into trouble.

Some literature suggests that it is almost a prerequisite to succeed in breeding this species, but I am not convinced of that. The creation of a natural watercourse with a small fall is very tempting from an esthetic perspective.

I built a watercourse with a fall and two tiers plus an overflow into an adjacent lily pond on the outside the aviary. The pekin robins loved it, but disaster struck when their clutch of two chicks fledged and the larger one was found dead in the small, shallow pool, about three inches (7.6 cm) deep. Presumably the chick was attracted by the moving water, entered it, and became confused and overcome by the chill of the water.

Since then the deep birdbaths in the outdoor aviaries are filled with pebbles until the chicks become more independent.

3.3.14 Lighting

The number of daylight hours a bird is exposed to affects breeding, migration, and feeding behaviors. An increase in daylight hours (photoperiod) stimulates nesting behavior in birds that migrate or remain in latitudes with marked seasonal changes in day length. It appears that twelve or more hours of daylight, combined with other factors, triggers the readiness to nest (chapter 6.2). (However, for birds that live in tropical and subtropical regions, where the amount of daylight is more constant throughout the year due to proximity to the equator, the breeding period is governed by the onset of the rainy season, which generates a rising food supply.)

In the wild (*in situ*), insectivorous birds in cold and temperate climates migrate to warmer regions, following the presence of insects and the opportunity to feed for more hours per day. These birds, having a high metabolic rate, need longer days to ingest enough food to last them through the night.

Some literature suggests that artificial light, activated by timers, should be provided during a short-day season for insectivorous birds kept in human environments (*ex situ*). The aim is to extend the daylight to ten hours, thereby giving the birds more time to feed.

Despite this, I have found that in our region (latitude 49.3°N), some birds still prefer to sleep in their dark, outside enclosures and thus do not receive enough light to feed for more than eight hours. Although additional light is provided, giving the birds the option of staying inside in a well-lit area while it is dark outside, most happily accept the short days and roost in the dark for longer nights. These birds are, of course, in good condition and live through the seasonal changes towards the short-day period by building up fat reserves. They are fed a high-calorie diet, including an egg cake (chapter 8.11), which helps sustain them over the longer nights. It must be emphasized, however, that diligent feeding is critical during the cold weather season.

I still recommend extending the length of day for birds during both acclimation and the molt, very young and old birds, and those in poor condition.

3.3.15 Live plants

Live plants are important to establish an aesthetically pleasing display, the mental and physical wellbeing of the birds, the provision of nest sites, breeding success, creation of microclimates, shading, prevention of predator injury, and photography.

Some of the birds like to build their nests in soft-needled conifers such as cedar or fir, but most prefer dense bamboo. These plants can be purchased as hedge plants and potted plants. Since most nests are placed between three and six feet above ground, a "nesting tree" can be bought and installed instantly. One tree or plant alone is not enough. The combination of two to three plants such as bamboo and conifers or deciduous bushes create an attractive habitat for both the birds and the viewer.

Figure 3.3.15-1 Irrigation sprayers should not reach the nest sites.

For an indoor aviary with normal room temperature, more tropical plants should be selected to cope with lower light levels and the "seasonless" constant climatic condition. I have not explored the best plant species nor the best regime to breed pekin robins inside a home or a small, indoor, tropical aviary. Consistent results have been obtained by breeding pekin robins in outdoor aviaries with a naturally governed and defined breeding season. Without that opportunity, I would surely work on indoor breeding with effective habitat development and suitable plant selection.

Poisonous plants are a concern, particularly in small spaces, which may lead the pekin robins to pick on plant material. In my large aviaries poisonous plants, such as narcissus, laurel, rhododendron and azaleas, have been ignored. Still, it is wise to study the list of potentially poisonous plants and choose other species. If you house finches and hookbills this is a much greater concern.

Watering the plants is accomplished by installing drip irrigation, especially for potted plants. Outside the nesting season, plants in containers and in indoor aviaries can be taken out and washed off or switched with other plants to give them exposure to the direct UV rays of the sun, which has a sterilizing effect.

The maintenance of outdoor aviaries for breeding pairs of softbills is not problematic due to the low impact of housing only one pair and the young in a relatively spacious environment. Where the plants are in a substrate of natural soil the fecal material is, to a large extent, absorbed. There will be a higher concentration of feces below the favored perches and roost spots. It only takes a short time to discover the movement patterns in order to make adjustments by placing sand in these locations for easy cleaning. Pekin robins do not damage plants very much and pruning will become necessary after some time to maintain a balance between vegetation cover and open space.

Silver-eared mesias, on the other hand, regularly engage in leaf pruning and love to nip off passion wine and young bamboo leaves. This is more the case during the winter, when they spend more time in the indoor holding aviaries, rather than during their busy breeding season when they are outside.

3.3.16 Suitable aviary plants

The selection of suitable plant species or their cultivars is governed by the local climate (for an outdoor aviary), available space, toxicity, personal preference and suitability. Nurseries and horticultural references provide information on regional climatic zones, based on minimum temperatures and other factors that limit the survival of outdoor plants. Outdoor aviaries provide favorable microclimates, particularly if they are partially covered with translucent material during the cold season, thereby expanding the otherwise limited list of hardy plant species. Plant selection for temperature-controlled indoor aviaries is not affected by low temperature, but often by relatively lower light levels. Cold-climate plants, such as northern species of conifers, generally require more light than tropical plants, which limits their use for most indoor aviaries.

Any aviary can be provided with suitable plants, whether they are planted in the ground or in containers. Most plants that grow tall can be pruned to control their upward reach. Pruning leads to desirable dense growth for nest sites, however, tall growing varieties of bamboo, for example, giant timber bamboo *Phyllostachys bambusoides* would become unmanageable in small spaces with low roofs.

Plant toxicity is not a major concern for softbills since these birds do not gnaw on and consume plants to the extent that psittacines and finches do. Azaleas, rhododendron, and even narcissus, were growing in my first walk-through aviary for several years without any known effects of poisoning to the birds, which included European finches.

Plant selection may be influenced by thematic development of an aviary based on biogeography. For example, Southeast Asian plant species would be appropriate for pekin robins and other birds of the region.

The most important aspect of planting is the creation of particular zones to provide opportunities for exercise, roosting, nesting, seclusion, climbing (laddering) and open space for surveillance.

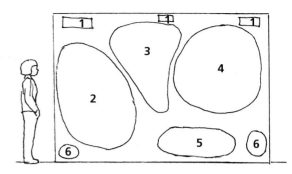

Figure 3.3.16-1 Planting zones of an aviary.

Zone
1. roosting and high perch zone
2. perching and feeding station zone
3. open flight zone
4. nesting zone
5. laddering zone
6. open space around perimeter

The space directly under the ceiling could be called the roosting/high-perch zone, where the birds have opportunity for long distance flight and resting on high perches. This area should be mostly free of plant material.

The perching and feeding station zone is where the birds approach the food and drinking water. Defoliated branches allow for easy approaches by the birds and a "window" for the keeper to look trough to the back of the aviary.

An open flight zone should be set up to entice birds to fly from cover to cover for exercise, and to provide opportunities to observe them.

The nesting zone is most critical for breeding. It must have dense plant growth to allow the nest to be hidden and well-anchored in the branch structure. Conifers with branches that grow in a whorl pattern (fir spp. genus *Abies*) are particularly suited for nest anchoring. I suggest using soft-needled conifers; they are nicer to work around and are probably more comfortable for the birds a well. I also use a lot of local red cedar *Thuja plicata* which provides excellent cover. Various species of bamboos, excluding the running ground cover type and the very tall, are ideal for creating nesting habitat.

There should be a laddering zone to allow fledged chicks to work their way up into higher branches for roosting. This is accomplished by keeping the lower branches on the plants. There is an advantage to having no foliage or needles on the lower branches; it not only helps the chicks to grasp branches, it also allows a better view under the vegetation to see what is on the ground. This is important for monitoring the progress of chicks, which usually start out from the ground up after fledging.

Finally there should be an open space zone around the perimeter to allow for maintenance and searching for unaccounted birds.

Figure 3.3.16-2
Western red cedar *Thuja plicata*.

Figure 3.3.16-3
Douglas fir *Pseudotsuga menziesii*.

Figure 4.1-1 Two ways to hold a small bird.

General Care

4.1 Capturing

Stress management is most important for this lively, fast moving, and alert species. Capture with nets is quite difficult, and complicated by the pekin robin's ability to learn not to be captured. Capture must be planned in advance and accomplished as quickly and safely as possible.

Capturing pekin robins in a planted aviary is practically impossible, unless you resort to the barbaric method of soaking the birds with a high-pressure water hose. If the aviary is connected to a service corridor the birds are best let out into that space. It is a good idea to let the birds out into the service area from time to time to build their confidence. Pekin robins are incredibly curious birds and love to investigate a new space that might yield a spider or some other insect.

Set up a perch in front of the open gate to make it easy for the birds to venture out. A few mealworms tossed on the ground helps the process. You may have to use a pull-string to close their door once they have come out, since they may return to their familiar aviary before you can get to it.

Once the birds are in the service area they should be moved into a smaller space that has few obstructions. A net curtain at one end of the corridor is most practical. The birds can then be caught with a hand net.

Pekin robin-size birds can be held in two ways (figure 4.1-1). It is important to fold up the wings and to position the legs in proper body alignment to avoid sprains and injury to the limbs. Be careful not to hold the bird too firmly as this can cause severe stress, and interfere with respiration and heart function. And if the bird manages to slip from your grip do not grasp the tail or you will find it in your hands. This happened to me when I pointed a pekin robin's head into a shipping crate's entry hole and the bird turned back, trying to slip out of my hand. It succeeded — or

Figure 4.1-2 A roll-up/drop net curtain aids in capturing a bird in service corridor.

Figure 4.1-3 Catching pekin robins by the tail is not effective.

rather most of it did. It takes about four to five weeks for a bird to re-grow its tail lost due to fright molt (chapter 2.8).

4.2 Trapping and Handling

Rather than chasing them down in an unsuitable space, it is far better to trap the birds in a trap cage and coax them by driving them into a small sock-shaped net that is attached to a trap cage (chapter 3.2.1).

The trap cage is light and small for ease of handling. The key features are: an extra track to receive the net, hooks to hang the cage on the aviary walls, a trip perch so the bird can catch itself, and fly screen sides. Fly screen walls are necessary to keep the waxworms and mealworms in the trap and, most importantly, to protect the bird from injury. Metal fly screen, not plastic, must be used to stop waxworms and mice (accidentally trapped) from chewing through it and tits from pecking their way out in only a few minutes.

A slide-out floor is very useful for cleaning, or for removing birds from an air-shipping crate with several compartments. In the latter case the lid of the air-shipping crate is shifted to create an opening for the birds to fly up into the trap cage. The trap cage may need to be covered with a cloth to keep the bird(s) quiet.

After some design evolution, this trap has become a mainstay for me to catch and transfer all my birds, including various finches. The trap cage is also suitable for transporting birds for short trips, but not by commercial carrier. Besides, once you try this trap cage, you will not want to let go of it.

The trap cage has vertical slides on each end. One slide is set up to drop when the bird lands on the trip-perch. The weight of

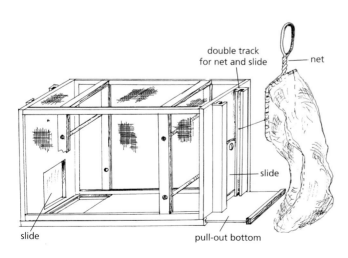

Figure 4.2-1 A trap cage has double track to slip a sock net in place. It is covered in fly screen to help contain bait insects.

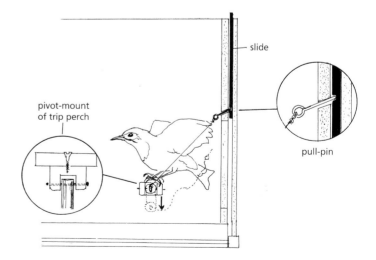

Figure 4.2-2 A hinged trip-perch pulls a pin to release the trap door when the bird alights on it.

the bird pulls a pin that holds up the slide. This works well for random trapping and does not require anyone to watch the trap.

The trap can also be operated by hand if there are several birds in the aviary and you need to catch one in particular. Either slide can be dropped with a pull string, using the following procedure.

- Bait the trap with mealworms or waxworms and set it on the floor, or hang it on the wall. Raise one slide and set up the pull pin with the string attached. When the desired bird is inside the trap, pull the pin to drop the slide.
  It sounds easy, but it is not. As noted earlier, Pekin robins "grab and run". If the bird has a mealworm already picked up, it will likely be out before you can block its path. It is best to wait for it to return and pull the string when it is on the way in, or if it hops up on the high perch.

- Once the pull pin is tripped, slide the sock net into place, cover the trap with a dark cloth, and pull up the slide in front of the net. Often the bird rushes in immediately. Drop the slide quickly so the bird can't return to the trap cage. Because of the length of the net, which entangles the bird, it rarely comes back out.

- Close off the net by grasping it near the opening and lift it out of the track. You can now reach into the net, fold the bird's wings into your hand and remove it from the net. Alternatively, you can leave the bird in the net and transfer it directly to another aviary, cage, or shipping box without the need for restraint, or simply carry the trap cage with the bird inside to the new location.

A warning, even at the risk of being repetitious: do not build a trap out of mesh wire with openings greater than 1/8-inch (3 mm) or, heaven forbid, convert a budgie cage with horizontal bars into a trap! The frantic birds will skin their forehead and faces, and, if left for a long period of time, severely injure themselves. Birdcages with vertically spaced bars cause less facial damage but, nevertheless, should be avoided for most small softbills.

It makes sense to place the trap box into the aviary ahead of time to allow the birds to become familiar with it. This pays dividends when a bird gets out into the garden. A single escapee will try to rejoin the ones left behind and the trap cage speeds up recapture. If a pair gets out, hunger will be the best incentive to lure

Figure 4.2-3 Two trap cages side by side with one bird already trapped lures another cautious bird into the open trap.

them into the trap. Once one is caught the other will stay close by, to be trapped later.

If you are dealing with a very shy and cautious bird, then there is great merit in having a second trap cage, or a cage matched in size, and placing them next to each other. One cage can hold the companion bird and the other can be ready to trap the bird on the loose. Many birds have been recaptured when they return to their faithfully used roosting perch at nightfall. In any event, it is senseless to run about wielding a net and causing a commotion, when the birds are better left alone to come home on their own. If there are predators around to disturb the birds the recapture is more troubling.

The shift boxes mentioned in the section on housing (chapter 3.3.6) can function in comparable ways. A pull string trips the slides. Standardize the shift boxes so that one slide on the trap cage lines up with that of the shift box. By placing the two units end to end the bird can be moved into the trap cage and netted as previously described without stress and risk of escape.

4.3 Predator Management

Predators come in many shapes and forms to harass, stalk, and attack our birds. Aside from parasites, which we will address in chapter 11 (Health Care), predators can include wasps that compete for live food in the self-feeder (chapter 6.10), blood-sucking mosquitoes that may transmit avian pox and West Nile virus, slugs and earthworms that act as intermediate hosts contaminate the aviary environment with nasty parasites; and there are the much larger uninvited creatures we must control. Rodents, raccoons, and other animals bring infectious diseases to birds. Even "harmless" birds become vectors for diseases and can put our aviary birds in jeopardy.

Prevention of break-ins and deterrents that keep predators away from the facilities is worth every effort. Time is well spent in studying the biology and behavior of predators in order to manage them.

In this chapter we will concentrate on predators that may attack our birds directly.

4.3.1 Rodents

The unexplained disappearance of birds, particularly young ones, from aviaries is often connected to rodent invasion. Rodents, as do other animals, have a greater need for protein during their breeding season, which increases their interest in animal protein sources. Food supplies and feeding stations will attract rodents .

The widespread Norway rat *Rattus norwegicus* and the black

rat *Rattus rattus* which has more southern distribution in North America, are the main culprits. Rats will chew through wooden walls and plastic netting without hesitation to get to the birds or their nests. Diligent construction and constant surveillance is the only defense. Trapping may reduce their numbers, but will not eliminate the break-ins. Heavy mesh wire curtains buried around the aviary perimeter, mesh wire floors, and concrete or other stone barriers will keep them out (chapter 3.3.3).

The much smaller deer mouse *Peromyscus maniculatus* which weighs almost the same as a pekin robin, would not attack a full-grown bird, but could become a problem with nestlings. On one occasion I discovered where deer mice had fed on a dead chick. The chick had fledged days earlier but had been doing poorly and was unable to fly up onto a low perch or into the vegetation. The chick may have died on its own, but it is conceivable that the mice cornered and killed the weak bird. Mice were trapped the next night baited with remains. Deer mice can climb like squirrels and find entry from the roof and high up on the walls.

Deer mice have more carnivorous tendencies than the smaller common house mouse *Mus musculus*. Deer mice feed on spiders, caterpillars, centipedes, and other insects as well as seeds. Any rodent should be kept out of the aviary as they will contaminate and consume the birds' food, and transmit disease.

Figure 4.3.1-1 Norway rat.

Figure 4.3.1-2 Deer mouse.

4.3.2. Trapping rodents

Trapping rats or mice must be done with caution. A pair of pekins will soon discover a forgotten trap in the service area, if they are let out to be captured or moved. The results need not be explained. To avoid such unforgivable events, either live traps or safe bait boxes should be used.

A bait box can be built which only allows mice to enter and be caught by standard spring traps inside. The safest design only permits access through a hole in the floor of the trap box. Two strips of wood raise the floor of the trap ¾-inch (19 mm) off the ground, allowing mice to make their way under it and through the hole in the bottom to the floor where the traps are set. The traps are baited with long-lasting peanut butter. The trapped mice are taken out through a removable side.

Often the mice concentrate on the food dishes, which are left overnight, and ignore set traps. The food dishes can be removed, however, this means the birds can't feed until they are served the next day. This is where the slide-out feed tray comes in handy.

The trap is set after dark and checked as early as possible to give the birds as much time to feed as possible, especially in the winter. Pull out the tray and angle it until it creates a ½-inch (12 mm)-wide — not a ¾-inch (19 mm)-wide — gap between the back

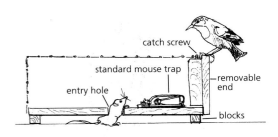

Figure 4.3.2-1 A trap box allows mice to reach the set traps inside by entering from the bottom, but does not allow the birds to enter.

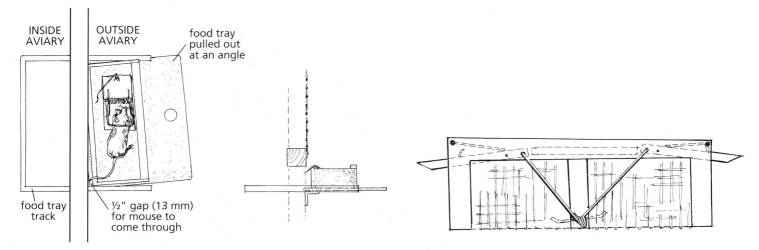

Figure 4.3.2-2 Food tray is pulled at an angle to create a gap for mice to reach the trap outside the aviary.

Figure 4.3.2-3 A Havahart live trap.

wall of the tray and the frame. Remove the food dishes and set a baited spring trap or live trap in the tray to catch the mouse. You can't forget this trap, because you will be feeding the birds the next morning.

There are different live traps, and some work better than others. A small Havahart live trap is a good investment. To the rodent it looks like a run-through. If it is positioned in the right place it hardly needs baiting.

Some might suggest using a pail with a wire stretched across and a baited yarn roll threaded on it. This works, but do not even think of filling the bottom with water. It will be as deadly for a bird as a rodent. In fact, do not store water in watering cans or steep-sided, open containers in and around where your birds could possibly have access to them, to avoid losses through drowning.

4.3.3 Other small predators

The disappearance of hatchlings may also be due to a garter snake *Thamnophis* spp. where they occur. I have no proof of them eating chicks, however, I have seen them climbing very near nest sites without being scolded or attacked by the parent birds, which were off the nest watching the snake. While the snakes are likely searching for tree frogs, they may discover naked chicks and possibly consume them. If the parents remove nestlings they can be found on the ground below and along the perimeter of the aviary in an attempt to take them as far away as possible from the nest site. If a snake has been in the aviary and the hatchlings disappear without a trace, the snake should be suspect. Garter snakes can't be kept out with ½- by ½-inch (12 mm) mesh wire. Unless smaller dimension mesh is installed one must watch for snakes and remove

Figure 4.3.3-1 Garter snake.

them. No doubt, other species of snakes are of concern in other regions and countries.

Mink *Mustela vison* do devastating damage to domestic fowl and aviary birds if they get access to them. They are not great diggers, but they are crafty in searching for openings that could be widened. A patch of rusty, weakened wire would be a concern. Unless interrupted, they will kill every bird present and thus can be a real problem if they enter the aviary.

Figure 4.3.3-2 Mink.

A very small, but fearless predator is the shrew *Sorex* sp. Due to its voracious appetite and tenacity it could potentially attack a nest or kill a small song bird. The animal lives in solitary and is rarely seen because of its nocturnal habits. More likely seen in rural areas, shrews are found in the temperate and northern climate zones of most continents. You may find one accidentally caught in a mousetrap or see one scurry across the floor in their typical back and forth running pattern of hunting for food.

Figure 4.3.3-3 Shrew.

4.3.4 Cats and raccoons
Predators trying to seize birds from the outside are a concern.

Cats are the most common candidate to harass and injure birds, even though they may not get access to the inside of the aviary. Injuries from cats generally occur during the night when they are most active. For the most part, un-neutered tomcats show up on their reconnaissance for females in estrus.

Pekin robins sit rather stolidly on the night roost and can remain undetected by a visiting cat. However, problems arise when the cat gets onto the roof too close to the roost site and begins chasing the birds from one end of the aviary to the other. Once the birds have bolted the cat will pursue them relentlessly until they are either exhausted enough to stay on the ground away from the mesh walls or, unable to see and perch again, they crash and land on the mesh wire where they freeze in fear, only to be hit again by the cat which can see perfectly well. The birds can incur serious injuries similar to those suffered by birds held in budgerigar cages and small mesh wire covered containers.

You may be awakened by the sound of birds slamming into the wire and the cat's claws raking up and down the aviary. If this goes on long enough, in the morning you will find the birds with bloody, facial injuries or even broken necks.

It is desirable to have a horseshoe-shaped, solid barrier at the back of the aviary to provide refuge for the birds. A night-light is also recommended to help the birds to find a safe perch again. An otherwise pretty, hexagon gazebo-type aviary, with mesh wire all around should only be used if the birds are locked away at night.

The cat usually returns to repeat the entertainment and a trap must be set to remove the feline once and for all. The local SPCA

Figure 4.3.4-1 Cats like to investigate dark places.

Figure 4.3.4-2 Raccoon.

(Society for the Prevention of Cruelty to Animals) or another animal shelter should be the next step for help. These organizations may lend you a live trap as well. It is best to bait the trap with catnip rather than cat food or you will catch nosy, free-roaming birds time after time. Food may offer little incentive to the cat since it may have just had a good evening feed at home.

Placing the trap effectively is important. The trap is best placed along the side of the aviary with pieces of plywood leaning up against the wall to create a form of tunnel leading to the trap. Cats love to check out dark, tunnel-like places.

The raccoon *Procyon lotor* is another predator to watch out for. This animal's ability to open doors and slides rivals that of a monkey. Loose wire soon becomes an opening to get inside the aviary. If the birds manage to get away from a raccoon, they might be able to escape through the sizable opening it created for entry.

A persistent raccoon can be trapped in the same way as a cat. People have used bananas and even chocolate for bait, but peanut butter and marshmallows are the standby. If you have your expensive softbill food stored in an insecure place, you will find much of it missing after a raccoon's visit.

4.3.5 Birds of prey

Birds of prey are another concern. For example, in the North American region the short-winged sharp-shinned hawk *Accipiter striatus* and the Cooper's hawk *Accipiter cooperii* are agile woodland hunters and launch surprise attacks out of nowhere. Usually the alert Pekin robins retreat in time to seek cover; plenty of vegetation cover in the aviary allows the birds to hide until the predator has left.

Figure 4.3.5-1 Sharp-shinned hawk striking at an exhausted bird.

Figure 4.3.5-2 Cooper's hawk.

However, panic crashes into the wire are problematic, and a bare aviary with only a few perches and less than two solid a walls becomes a death trap. The hawk will circle and attack the panicked bird hanging onto the wire, attempting to remain motionless. Conceivably, the hawk's talons could actually reach through the mesh wire and injure or kill the bird inside.

Unfortunately the hawks are not easily convinced that they can't take a bird out and make many repeated visits to try again. This harassment may cause all the birds to remain in the inside enclosures or in dense cover for days. Most of the singing and nesting may stop.

There is little you can do, other than drape greenhouse shading cloth or similar material over the most exposed sides of the aviary to reduce the hawk's impulse to dive for the birds inside. Failure to seize a bird will eventually cause the aerial attacks to cease.

Owls can be a concern if the roof does not have solid coverings. The most likely loss would occur in the dark, if an owl causes the birds to leave their roost and fly to the wire.

Night lighting helps the birds to avoid panic, to find their perches, and to better see predators. If nocturnal predators become a problem and the aviary does not offer suitable roost places, then such lights should be installed. Night-lights are regulated by photocells for convenience and saving energy.

Figure 4.3.5-3 Owl attacking a sleeping bird.

4.4 Flock Management

During the winter, wild pekin robins band together in flocks to roam the countryside, often joined by other species. Pekin robins travel with other babblers, such as blue-winged sivas *Minla cyanouroptera* and red-tailed minlas *Minla ignotincta* (chapter 12.4), from higher mountain elevations and northern regions to warmer climates during the cold season.

Figure 4.4-1 A flock of babblers.

These gregarious tendencies are also noted in our aviaries. In particular, young birds from the current year, and unmated adults can safely be kept in a flock situation for three to five months (generally until September to November) after the end of the breeding season. Pairing up of individuals commences long before spring. As aggression among the birds erupts with pair formation, I generally isolate the pairs to their own breeding aviary by the end of November.

It is not unusual to see persistent quarrelling and fighting building up between adult birds well outside the breeding season. If this occurs the incompatible birds must be separated to prevent injury or even death. Pekin robins can and will kill others in close quarters, but also in large planted aviaries (chapter 6.5).

To avoid the formation of sibling pairs, more than one aviary is needed to separate the males from their sibling females (chapter 2.6). You may also see bonded pairs form between same-sex birds as part of normal social behavior.

Mixing new pairs with a pair left in their established breeding territory may not be successful even in the off-breeding season. A neutral and large aviary with good plant cover is a much better place to house established pairs in a flock situation during the winter.

There is actually little advantage in putting pairs into a flock if they are later returned to their own spaces and their aviaries are kept empty in the meantime. Saving time by serving fewer food stations is outweighed by the problem of not being able to monitor individual food intake and condition of the birds or, most importantly, sudden, serious fighting.

4.5 Observation

The observation and study of aviary birds is fundamental to their successful management. Ethology — the science of animal behavior — is built on observing animals. Maintaining birds in artificial environments places the burden of responsibility for their welfare squarely on our shoulders. Daily observations become baseline information about their needs and desires, and how well they are met.

Highly spirited insectivores are hunters that react immediately to your arrival and change their behavior accordingly. If you remain motionless, without holding promising mealworm containers in your hand, their activity will "normalize" and other priorities gain their attention. The longer you remain that way the more you will discover.

It is quite acceptable to be a "peeping tom" around your birds. I use small holes in strategic places created for this purpose, watch them from darker places through glass, and resort to the ultimate tool, a security camera. An inexpensive, outdoor, color camera, sold for monitoring events around your home, can be mounted near a nest site, in the quarantine unit, and in other places. A TV monitor may be a long distance away in another building or office. These units are inexpensive and come equipped with audio input.

I cannot express the relief this tool has given me since losing a mesia chick, which was tossed out by its parents and succumbed on the ground unnoticed, after closed-banding. Now parents at the nest are on television, in sound and living color, long before the chicks are banded. Before this, I could not check on a pair's feeding behavior, because they simply stopped and watched me until I had obviously walked away. I am convinced that both the birds and I will live longer with this ingenious stress reducer.

A note pad is standard equipment near the monitor. To economize the time spent on observations, subsequent nesting events can be predicted by reviewing previously recorded activities.

Important kinds of observations and information to record include the following:

- inventory changes
- health issues and treatments
- breeding behavior (courtship to weaning of offspring)
- other behavior
- body weight (when obtainable)
- construction and repair of facility
- environmental changes (weather)
- diet changes

A security camera is a fine behavior study instrument and helps to share the joy of watching the bird family with visitors at a distance away from the nest site.

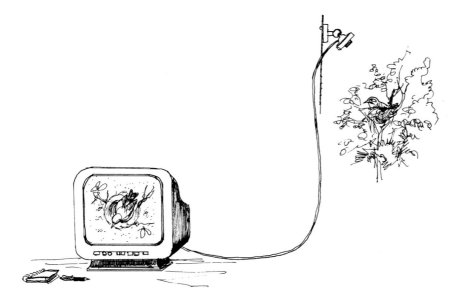

Figure 4.5-1 A security camera transmits the events at a nest to a monitor.

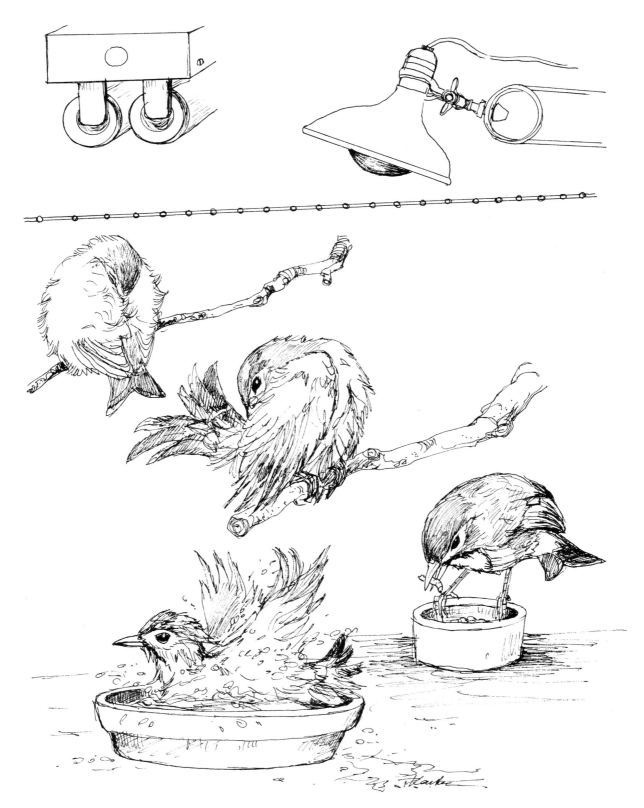

Figure 5.0-1 New arrivals need light, warmth, food, a bath, and an undisturbed environment to find food, conserve energy, clean up, and rest to recover from shipping stress.

Acquisition, Quarantine, and Acclimation

5.1 Acquisition

While it might seem to make sense to present the chapter on how to obtain pekin robins at the beginning of the book, it cannot be stressed enough that the aviary, quarantine room, and food supply must be set up before rushing out to acquire birds. The chapters on biology, housing, and general care have been presented first specifically to emphasize that point.

When the commercial importation of this species was allowed, obtaining a pekin robin was simple. Some pet shops carried them regularly until 1998. The species is still popular and very much sought after despite its unavailability in the pet market. Fortunately there are avid aviculturists who breed pekin robins and other softbill species. To locate them one can search the Internet and contact local or national bird clubs and avicultural societies for leads.

Serious breeders band the chicks they breed with closed, numbered, and dated leg bands that are issued by national avicultural societies. In some instances studbooks are set up to genetically manage an international, national or regional population among a consortium of breeders. The studbook keeper can assist in making contacts to gain access to birds that are otherwise difficult to find. Avicultural journals and cage bird magazines will place advertisements to search for birds and birds for sale.

For the most part the reader will have to depend on the availability of single birds or pairs that can be located from breeders. A dialogue with the breeder or owner is critical to establish the status of the bird(s) and to gain insights for making decisions on acquisitions. When possible, it is always better to visit the facility; the benefit of seeing a bird first is invaluable.

The now rare opportunity may arise where one or more birds can be selected out of a flock of several birds that are offered for sale. In that situation a set of rules can be followed for choosing good, strong birds of likely good health. If there is no choice given, these points of observation are still important in order to assess the health and condition of any bird.

5.2 Importation

The international trade agreement of endangered species (CITES) now restricts the importation of pekin robins, and many other listed species, to authenticated, aviary-bred birds, and in much lower numbers than in the past. However, for the many softbills not covered under CITES, exportation still begins with capture in the wild and transfer to local bird markets.

Exportation from the originating country and importation to the recipient nation is an arduous process for the birds (whether aviary-bred or wild-caught) and the involved agencies. The highly stressful episodes involved in international importation — transfer from the breeder or bird markets, storage in the exporter's facility, then international transport, federally imposed quarantine of twenty-eight or more days in crowded quarters, and further overland travel to a pet shop or private buyer — is an ordeal only the strongest can survive. Crowded environments challenge the immune system of any bird.

It does not end there. Potential exposure to other unhealthy birds and pathogens, and abrupt changes in environmental temperatures and unfamiliar or even unsuitable diets, are constant threats that weaken the bird. Ailments and health problems might be expected with new arrivals.

Bird importers have developed better methods of shipping and acclimatizing birds to minimize losses. New imports are important to inject new founders, which are genetically unrepresented contributors to a breeding group of animals (gene pool), to avoid inbreeding when unrelated specimens cannot be located.

Hopefully more consistent *ex situ* propagation of aviary birds will become more commonplace and reduce the need for importation from other countries. In 2005, out-breaks of highly pathogenic strains of avian influenza have had a dramatic impact on imports and exports of birds as well as the movement of same within affected countries. Contagious diseases are of great concern to aviculturists and newly acquired birds must always be quarantined, whether they come from another country or a neighbor (chapter 5.7).

5.3 Obtaining Background Information

Often a bird is located too far away to inspect it. Here are some points on how to evaluate a bird and what questions to ask:

- When was it hatched?
- Does it have a closed, registered, dated leg band?
- Does it have any ID, color leg band, or unique physical features?
- Is it part of a studbook program? (parentage)
- Where was it obtained?
- What is the estimated age or hatch date?
- Has it bred before?
- Does it display aberrant behavior (e.g. killing other birds, nesting problems, highly aggressive)?
- Is the sex known with certainty? (DNA test?)
- What is the medical history? Any previous health problems?
- Why is it for sale?
- Will the bird be guaranteed for live arrival, and for what period of time thereafter?
- Are photos of specimen or facility available?

If a visit to the facility to inspect the bird is possible, the above questions are still relevant, however, direct examination also provides further information.

5.4 Judging a Bird's Condition

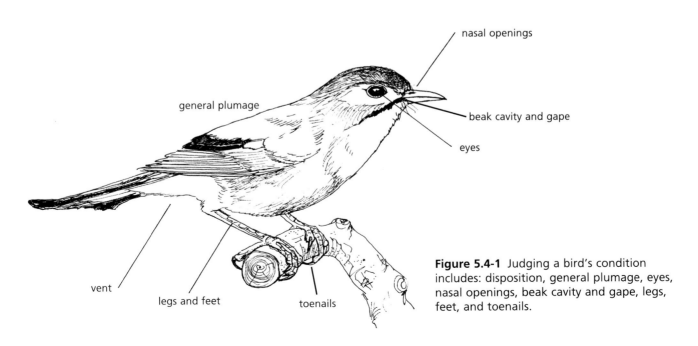

Figure 5.4-1 Judging a bird's condition includes: disposition, general plumage, eyes, nasal openings, beak cavity and gape, legs, feet, and toenails.

Figure 5.4.1-1 Damaged plumage on new arrivals is not necessarily a sign of poor health.

5.4.1 Plumage

The first feature we notice is usually the plumage; but the bird with the best plumage is not necessarily the healthiest. Good plumage and the right behavior must go hand in hand. A number of damaged wing or tail feathers does not necessarily indicate a sick bird.

I can recall an episode where, in attempting to release a pekin robin into a shipping box (chapter 4.1), I accidentally transformed it into what looked more like a blue-breasted quail when the tail came off in my hand. Few buyers would be happy to acquire a tailless bird, however, the recipient was advised of the mishap and the tail was fully replaced to its former splendour within four weeks (fright molt, chapter 2.8).

Bald spots on the body may be a concern, as they could indicate feather lice or an abnormally progressing molt. In a pet shop with birds cramped together, however, feather picking is a common problem and some species are more prone to this than others. The red-tailed minla *Minla ignotincta* is notorious for this, and pekin robins are also candidates to de-feather their cage mates.

Case report
A case in point was a red-tailed minla the importer did not want to ship because it was "naked". Reportedly it was active and eating well. It looked rather bare when it arrived, but not for long. Paired up with a lone female it blossomed within a month to have picture perfect plumage. In full plumage the species is dimorphic, as the male is more colorful and can easily be sexed by appearance, which is not entirely the case with the monomorphic blue-winged siva *Minla cyanouroptera* or the pekin robin. However, upon introducing it to a hen it revealed its sex — in spite of its plumage condition — simply by its behavior in acting out an unforgettably comical, featherless courtship dance. It is good advice not to look for perfect plumage alone, but to observe what the bird does.

Be alarmed if the vent feathers are soiled; it could point to infection in the digestive system. Also, feathers in poor condition around the face could be due to a discharge from the nasal opening or eyes, an indication of a possible infection in the respiratory system. Unexplained exudates need further investigation (chapter 11.2.1).

5.4. 2 Feet and legs

Feet and legs should be clean and complete. At times we see scaly legs caused by an ecto-parasites. Older birds tend to have a rougher texture on their legs and feet. Swelling around foot joints

Figure 5.4.2-1 Rough scales may indicate the presence of parasites, high age, or poor husbandry.

may indicate renal dysfunction (chapter 11.2.4), a leg band that is too tight or arthritis.

A bird may be missing toes. The first hen I acquired, in 1999, has no hind toe and only two functioning front toes on one foot. A parrot was blamed for the injury. She looked like a poor prospect to ever raise a brood, but did just that a month after she was placed with a male, and continued for years afterwards. Imperfections due to old injuries are not always a problem for breeding. A damaged bill can be an issue, but with a little help it can reshape itself.

A cosmetic flaw is perhaps a concern to bird keepers who enter their birds into bird shows, but not for the breeder, provided that the imperfection is not genetically based. This may become a probability after multiple generations have been raised from limited founder stock. With many wild-caught birds currently in the gene pool of pekin robins, there should be low incidents of genetic inbreeding defects.

5.4.3 Disposition and body condition

Take time to observe the bird or birds you want to obtain. Behavior is the best indicator of hidden conditions. Do it at a distance. If you are within the flight distance of the flock, they will all show distress, especially the healthy ones. The problem is that all will display the same behavior. A sick, but threatened bird will mask its ailments masterfully. This is a predator avoidance tactic so as not to become a target of pursuit. You must back off and let their personality or state of well being present itself.

A healthy, well-adjusted pekin robin soon will show its natural curiosity in its surroundings, including you. It may even scold you, which is a good sign. A sick bird has no energy to move more than necessary and does not respond to your approach as much. Place a few mealworms into the enclosure and see the reaction of the flock. Unless they had their fill, you will see who is who in the hierarchy very quickly. Healthy birds should show keen interest in the offered mealworms.

It goes without saying that a bird sitting "big" with its feathers fluffed-up for extended periods of time is a concern. Likewise, a bird that sits at the food dish without actually feeding or, worse yet, sits on the floor with its chest and beak resting on the ground, is worrisome. Unlike softbills, a healthy finch or hookbill will stay at the dish and fill its crop, something most softbills do not have, although doves and pigeons which are grouped with softbills have a crop. It is not normal for a pekin robin to remain at the food dish for long.

Check body condition by taking the bird in hand and blowing its chest feathers apart to see the muscles on either side of the keel (breastbone). Ideally the chest should be filled out and neither

Figure 5.4.3-1 Sitting "big" and tail bobbing indicates discomfort.

Figure 5.4.3-2 Sitting at a dish without feeding is a concern.

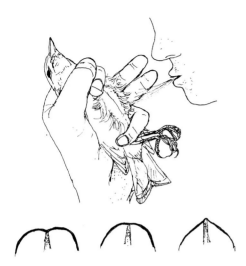

Figure 5.4.3-3 The chest is revealed by blowing the feathers apart to expose the chest's condition. The center profile is ideal.

Figure 5.4.4-1 Exudates from the eyes and nares cause wet and dirty facial feathers.

bulging (too fat) nor emaciated (bony). The latter indicates that the bird had insufficient food, or is fighting or recently suffered some illness. There may be known reasons for being undernourished, such as importation, or long quarantine and shipping, so this may not be a serious problem, as the bird will likely recover with good care.

An aviary-bred bird, which should have had access to plenty of food and low stress, should show good muscle development. Parasites and hidden ailments, which interfere with food absorption, cause a bird to lose its fat deposits and muscle mass. Every effort should be made to obtain a weight to compare it to the average weight of the species and to establish a base line (chapter 7.6).

5.4.4 Eyes, nares and beak

Birds with respiratory infections often have discharges from their eyes and nares (nasal openings). The eyes should be bright and wide open while the bird is active, feeding, or busy in its surroundings. Wet feathers due to exudates from the eyes and nares indicate the presence of a respiratory or related disorder. Great caution is advised when choosing a bird with these symptoms because the ensuing stress of moving the bird will aggravate any infection. The inner lining of the beak and mouth cavity should have healthy tissues. The feathering should be complete at the base (gape) of the beak. These are areas where fungal infections may be noticed in an unhealthy bird.

5.5 Supplier's Facility

The general cleanliness of the aviary, the diet presentation and the food itself, record keeping, and health of the entire flock, are important indicators of competent management. Overcrowding or marginal diets may be a reason for a bird in poor body condition, but much can be done, once it is in your care, to improve this. Filthy environments are sources of pathogens and one must guard against introducing them into the home aviary. This makes quarantine procedures all the more important. Obviously, sick birds in a supplier's inventory are a reason not to acquire any birds from that source.

5.6 Transport Home

Taking the bird(s) home should be done swiftly. Capture and confinement is traumatic and very stressful for active and not-yet-domesticated birds like pekin robins and most other softbills. A baited transport cage/crate can be placed in the aviary to get the bird(s) used to it (crate training) to reduce capture and shipping stress.[1] A short transport of a few hours can be done in a dark cardboard box with small or screened ventilation holes. The box should have a perch and some absorbent tissue on the floor; the frightened bird often gets diarrhea and it would be a shame to bring home a soiled bird, if it can be avoided. The bird will not and cannot feed in a dark box, therefore, if it is kept too long in this situation it will dehydrate and starve. In hot weather this is a real concern.

For longer than half-day trips, a transport cage is the better choice. This could be a sturdy cardboard box with a screened front to permit light to enter the space so the bird can see its food and water dishes. It is a good idea to include sliced oranges as a source of liquids, since water may spill out. You can also "float" a slice of orange in the water dish to reduce spilling. The addition of a slide door in an upper corner is of great help to get the bird safely in and back out. The front can be draped with a light cloth while the bird is carried around and removed again once the box is placed in a relatively quiet place, such as the back seat of the car, to give it some light.

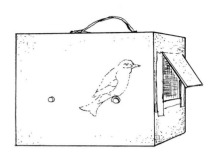

Figure 5.6-1 Example of a transport box.

Warning:
Never transport a pekin robin in budgerigar cage with the problematic, horizontal bars, spaced one-half inch (12 mm) apart!

Even a hand-raised pekin robin may not accept being caged up and moved around without a relentless fight. It will persist in trying to get out between the bars and skin its forehead in the process. The injury can take place in an hour or less, and is often so severe as to disfigure the bird for its lifetime, often causing permanent damage to the eyes.

We see this far too often, and all because of ignorant or careless handling. Fly screen is the best material for lining the inside of openings in small containers.

Although hundreds of finches and hook bills are shipped in wire-front cages with half-inch square openings, softbills definitely cannot be transported this way.

5.7 Quarantine

Strictly speaking, a quarantine space is an enclosure that isolates birds from other birds or animals that can be vectors for communicable disease. It must also prevent transfer of diseases via airflow and personnel. This demands certain facility design and operating procedures to prevent cross infection.

Bird quarantine periods are consistent with the duration of incubation times of diseases. To be on the safe side, quarantine should last four weeks, but you may make a judgment call to shorten this if you know the origin of the bird.

1. C. Sheppard, pers. com 2006.

After a few days of observation you may decide to place a single new arrival with its intended mate to give it comfort, and to help it adjust to the new diet. This means, of course, that you now must quarantine the two together for a correspondingly extended period, but it will speed up the mating and breeding schedule if the bird arrives late in the breeding season Should any infection become evident, both birds must be treated and cured before they are released to the aviaries where they will be in contact other birds.

To be effective your quarantine unit should really be in a separate building that is a good distance away from the resident bird population. (Federal quarantine units specify a 328-foot (100 m) separation). It should be an indoor only space to exclude contact with wild birds, insects, and rodents. Waste material, bedding, old food, drinking and bird bath water etc. should be collected in separate containers and stored until the quarantine has passed and no disease has manifested. The shipping box should be either safely disposed of or thoroughly disinfected with a 10 percent bleach/water solution. Contact a veterinarian if you suspect any disease.

A footbath at the entry porch to the quarantine unit is recommended, and changing into designated boots and coveralls only worn in the unit is strongly encouraged. The birds in the quaran-

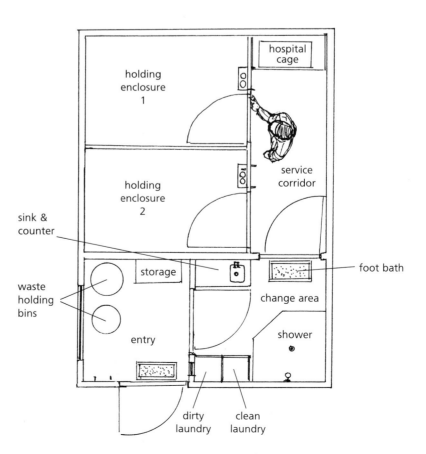

Figure 5.7-1 Layout of a quarantine unit.

tine unit should be served their food last, after all the other birds have been fed.

Biosecurity is a new concept in animal husbandry. Outbreaks in some parts of the world in 2005 of highly pathogenic avian influenza strains (HPAI–H5N1) are giving quarantine procedures a whole new meaning.

Should you decide to import birds from another country it must be done under permit by the authorities responsible for regulating livestock importation; and in the case of endangered birds, the authorities governing CITES regulations. A federally approved quarantine unit must be established beforehand. The unit must pass inspection prior to occupation, and prescribed operating procedures must be followed. These regulations are not exactly the same between countries and change from time to time within relevant jurisdictions; it is therefore best to make inquiries with the respective government agencies at the time you wish to pursue this. Government web sites are a good starting point to obtain current information.

It is a good practice to maintain a diary and consistent records of the birds, especially if medication is administered. The time to band a bird for identification and/or to record what closed-band ID it already has is when it enters the quarantine unit (chapter 6.14).

5.8 Acclimation

Acclimation or acclimatization, — either term can be used — should be standard procedure for all new arrivals. It is of particular importance for wild-caught or imported birds to condition them to aviary living. Imported birds are often in stressed condition and may have lost body weight to varying degrees. The plumage can be soiled and damaged to the point where it loses its insulating capability.

All this calls for high ambient temperature on arrival and over the next several days, at least in that part of the cage or aviary to which the birds retreat for greatest comfort and safety. The maximum temperature may be close to blood temperature (101°F/38°C) in some selected spots, but an average of about 85°F (29-30°C) is recommended. It helps with recovery and prevents body heat loss. Gradually the temperature is lowered to the normal values of the aviary the bird is moved to.

For the first twenty-four hours the lights must be kept on to allow the bird to find food and water. For new arrivals after air shipment this is a must since the birds may have been unable to feed properly during transport. Later, low wattage night-lights are adequate to give the bird a chance to perch again should it become disturbed.

Live food such as mealworms and waxworms are invaluable to entice the bird to begin to feed. It is helpful to obtain the diet that the bird ate prior to the transfer, but that is not always possible. This diet may be inferior but, in any event, it can be built on to improve it. It is best to offer a wide variety of food items including vitamins, amino acid, pro-biotics, mineral supplements, and electrolytes. Food consumption should be monitored: for a flock, direct observation is necessary; for a single bird, remaining food/water can reveal intake.

If illness is suspected, the water can be medicated or the bird can be treated directly. The provision of a drinking tube with sugar and/or vitamin mix placed near the food dishes entices the bird to drink from it rather than the birdbath.

A birdbath, which must be kept very clean, is needed to help the bird reconstitute the condition of its plumage. Usually birds are eager to bathe to clean up after shipment, but one must make sure that the environment is warm enough to avoid chills (hypothermia). Shipping may put the insulating structure and water-shedding capability of the feathers out of condition. This can result in soaking of the feathers and skin, and the bird becoming chilled and stressed (see case report, chapter 3.3.4). An infrared heat lamp can be set up to ensure the maintenance of body temperature until the plumage has dried again. It may take several days before the birds have reconditioned their plumage.

A bird received from a source within the country is less prone to acclimation problems. The ambient temperature at which the bird had been kept must be matched to avoid sudden changes. Still, to combat stress, it is wise to err on the high side during the days of transfer.

During the winter months, and even in the shoulder season, birds bred and raised in milder climates in an outdoor aviary cannot be transferred to an outdoor aviary in colder regions. Gradual exposure to changing seasonal conditions or varying conditions from indoor to outdoor environment is the key. To prepare softbills for living in outdoor aviaries means waiting until late spring or early summer to let them gradually adjust to the colder season.

Experienced aviculturists often hold new arrivals in relatively small cages to ensure that the bird will find food and water without difficulty.[1] Single caging allows for monitoring food intake, individual medicating of food, and the inspection of discharges. Ambient temperature can be controlled and so can activity around the cage. While a nervous bird will experience some stress, particularly when approached and looked at closely, it is by far less risky than releasing it into a big, planted aviary where it may perish unnoticed. Competition and aggression by other birds are further concerns in a community enclosure.

1. David Holmes, pers. com. 2004.

5.9 Release to Aviary

Once quarantine and adjustments to the new diets are completed, the birds can be released to their designated aviaries. This should be done early in the day to give the birds time to orient themselves, and most importantly, to find their food station and water supply. If these are provided in an attached indoor enclosure, it is necessary to introduce the birds to this space first. They should be held there until they are feeding well. The slide to the larger outdoor aviary should be opened in the morning to allow the birds plenty of time to explore at their own pace, and then return to feed inside.

A nervous bird may, nonetheless, decide to stay in the larger outside space and try to fend for itself by catching the odd insect. This is problematic and one may have to bring the bird back in for a day to feed properly, or set up a feeder in plain view outside.

Figure 5.9-1 For a new, timid bird it is better to bring the food dish to the bird than the bird to the dish.

Introduction to other birds needs to be monitored. While the others may lead it to the food, there may also be aggression and intimidation, which prevents the newcomer from feeding. A food dish placed near the hideout of the shy bird is helpful. The resident birds will be unfamiliar with it and, therefore, less possessive about it.

The behavior pattern for pekin robins that are introduced to each other resembles what is described under Mating (chapter 6.8). First contact between pekin robins usually results in a standoff posture by both birds (chapter 6.2). Compatible birds of the same or opposite sex will quickly sort out their hierarchical position by exhibiting dominance or submission. Tension generally eases off within hours and either pair or buddy relationships start to form, which is indicated by mutual preening. Rarely do we see persistent and serious aggression requiring intervention. Careful monitoring is still necessary for a few days.

During the short-day period we must give birds extra light hours for the introduction to the new environment; in fact, twenty-four light hours for the first day is highly recommended.

In very large aviaries an integrated introduction cage is helpful to let the new arrival gradually become accustomed to its new environment and cohorts. The introduction cage can also be employed as trap cage and secondary feeding station.

Figure 5.9-2 Soft release introduction cage.

Figure 6.0-1 Pekin robin feeding one week old chicks.

Breeding

Breeding birds embraces many elements beginning with creating and sustaining suitable environments, establishing compatible pairs in a breeding aviary, and providing proper nutrition and nest material; and then extends to monitoring nesting, laying and incubating eggs, hatching and feeding nestlings, and fledging and feeding fledged chicks until they become independent. The breeder's responsibility also includes artificial incubation and hand raising birds when necessary.

Breeding aviary birds is an increasingly important dimension of aviculture due to loss of species and their habitats, escalating health concerns, and endangered species legislation affecting the availability and movement of birds.

In 1997 pekin robins were listed in Appendix II of CITES, which prohibits commercial importation of wild-caught birds. Domestically raised birds require registered, closed, leg bands, plus import and export permits to move them from country to country. Having many softbill species as aviary birds for the future depends on our ability and commitment to breed them consistently for many generations. This chapter is written to turn occasional luck with breeding into understood and regular success.

Chapter two discussed how to sex birds (2.6). We will now look at other parameters, namely the determination of compatible partners, the nesting environment, nest support, diet adjustments, and specific daily management regimes.

6.1 Pair Behavior

The pekin robin is a contact species, which means the birds perch or roost body-to-body and engage in mutual preening (allopreening) with others of the same or opposite sex. Judging by their behavior, you might think that if you have two pekin robins you automatically have a breeding pair. Countless, convincing "breed-

Figure 6.1-1
White-rumped shama.

ing pairs" have been sold over the years. Two hens could even build a nest and have eggs. Because the species was usually captured in Asia for its singing and not for captive breeding, hens were not kept or exported. This is one reason why pekin robins have not been bred very often in the past. But even a true pair will not breed unless we provide the right environment and diet.

There are also birds, usually males, that can be aggressive. The species is very territorial by nature, and while this should primarily manifest during breeding season, it can extend throughout other seasons. This trait is inherent in the nature of many insectivores, which are dependent on a productive feeding territory that they must defend in order to secure enough food for themselves and their young. While raising a family, insectivores and birds feeding on live animal matter defend their territory more tenaciously than do seedeaters. Some species are difficult to manage in a common space. Even breeding pairs require careful monitoring of their behavior in order to avoid serious injury, or worse. A typical example among favored aviary birds is the white-rumped shama *Copsychus malabaricus* (chapter 12..14).

Pekin robins pair up easily and stay together all their lives. Switching mates only works if one is removed from sight and, in particular, from sound. Even at that, it can take many days or weeks before the new partner is accepted, despite the innate urge just to have contact as much as anything else. Even at the height of the breeding season it may take weeks before the birds will accept "forced marriage". Given the opportunity I let the birds choose their own mate.

A bird from a "bonded pair" of males will, on occasion, not accept a hen, which may even try to actively solicit contact, because she is regarded as an intruder. The surplus male must be moved from sight and earshot in order for the new pair to form. You may get the impression that the separated male, in ignoring the hen, has turned homosexual. But there is a loyalty complex between the two "buddy" birds that needs to wear off before the hen will be accepted. Two hens will form buddy relationships as well.

Sometimes a single male will not accept any hen presented to him, and will chase her viciously. In a cage or aviary without plant cover and escape routes, this can lead to the death of a hen. This aberrant behavior is not the norm and is probably triggered by unnatural conditions in captivity, in particular the lack of sight-barriers. Separating and isolating a bonded pair from each other for a period of time will build the desire to pair up and ease the introduction process; however, it seems some males remain incompatible and aggressive. This must be recognized in time to protect the other cage mates. Such males may be candidates for a mixed species show exhibit, but not for a breeding program.

One male was removed after he assisted the hen in nest building, but then repeatedly tore up the nest in order to build a new one, causing the eggs to tumble to the ground. The hen received another male and soon raised four chicks to weaning age before the close of the season.

The dysfunctional male showed extreme stereotypy when locked into a small cage, frantically repeating movements, including upward head twisting, over-preening, and other senseless movements. It can be assumed that the bird was psychologically traumatized by having been previously kept in small cages. Another male, with only a slight disorder of head twisting, lost his stereotypical behavior when he was moved to large aviary, but demonstrated it again if he was locked in a small cage.

6.2 Breeding Season and Reproductive Age

The breeding season is governed by the hours of daylight. Day length (photo period) is one factor among many which must be met. In my experience, at the west coast of Canada, the nesting the nesting activity of the colony roughly coincides with the vernal equinox, in the spring, and ends in September when the birds begin their fall molt. However, there are differences from pair to pair and bird to bird; and the season for each bird and pair can vary significantly. For example, some pairs are early starters with nesting activity commencing early April, but then do not breed beyond late August, while others breed from late May to late September. Some have a shorter overall season than others.

A pair of silver-eared mesias surprised me by laying their first eggs in late March, and hatching their last clutch in early September. They made seven nesting attempts in that season. This is an old, but newly acquired pair, which never successfully reared young before with the previous owner. Perhaps their new large, well-planted aviary released a lot of bottled up nesting desire.

The onset of the fall molt varies from bird to bird, with some consistency from year to year. A hen beginning her molt in August is likely to repeat this the following year under the same husbandry conditions. There is, in my observation, a relationship between the onset of the molt in parent birds and the "finishing-off" of the season's last clutch of chicks. The molt has never started before that event. In other words, the molt is delayed until the last chick is weaned.

While nesting activities may define the breeding season, the success of actually rearing chicks is not constant throughout the season. I experience a seasonal rising and declining curve, with a marked peak in July. In a seven-year study, half of eighty-three chicks raised were hatched in July (Breeding Statistics, 6.17). Chick losses are greater in the shoulder season (6.11).

On average it takes five weeks between hatching and weaning for the pair to rear their chicks to independence. But individual nesting periods can overlap, since the hen will usually lay eggs again and start incubation before weaning has taken place (6.13).

The natural breeding season may become erratic and not clearly pronounced in an indoor environment, with its artificial lighting and constant climate conditions. Breeding can, however, be manipulated by controlling lighting, and mimicking seasonal climate and dietary changes. The technique of manipulating daylength and temperature certainly works like clockwork in horticulture by forcing poinsettias to bloom exactly at Christmas time.

Theoretically we can control indoor breeding and mimic seasonal changes but, in my view, this is not a practical approach if outdoor breeding can be facilitated.

For a breeder it is important to note that a pair of pekin robins, moved in the fall from an outdoor environment to a heated, indoor space with extended hours of light each day, will likely begin to breed within weeks. However, it is not desirable for a young pair, hatched only five to six months earlier, to begin to breed at such a young age. I know of two pairs brought into a home situation that resulted in winter breeding. In these cases consecutive clutches were hatched in December to January, but all were lost before fledging. Besides being very disappointing, this puts a drain on the birds' resources and vitality. One of the hens was only six months old.

Breeding in human care gives us data on reproductive age and span of years a bird may breed. We know that young pekin robins may breed as early as six months of age and possibly sooner, as noted above. One bird fledged eleven chicks in four clutches in her first year of breeding, between the age of ten and fourteen months.

Aviary-bred pekin robins and mesias breed in their first year; in outdoor aviaries with a naturally arriving spring, they would be about eight to ten months of age. This may not be so in the wild. A harsh winter and lack of nutrition could influence sexual maturity. We also know of some bird species (gulls), which loaf around for their first year before they settle down to breed.

I have reproductive pekin robins in my inventory that are over eight years old. Since these are wild-caught birds they could even be considerably older. Presumably, pekin robins will breed well beyond ten years of age as long as they are in good health. Keeping studbooks will provide good information for the future.

Yearling hens are generally successful if they are breeding with an experienced, older male. Males, on the other hand, are less likely to successfully rear chicks in their first attempt. This will be discussed further under Incubation (6.9) and Feeding Chicks (6.10).

Physiological readiness to breed is but one of many factors in successful nesting.

6.3 Introduction of Mates

The best way to create breeding pairs is to let the birds choose their own mates. This is possible if you have several unrelated breeding pairs and allow their offspring to form unrelated pairs after they have become independent in their first fall. Siblings would be separated for this. It is, however, unlikely to be in that position and mates must be brought together.

One way to ready birds for pairing is by keeping them alone for a few days and, as noted above, totally separated by sound and sight from their previous companion. Hens can be reluctant to accept a new male. The best way to overcome this is to put the two birds in an isolated aviary, alone and away from other pekin robins.

It is also always better to introduce the male to the hen in her familiar enclosure rather than vice versa. The male is the more aggressive, and the hen is better off knowing all the escape routes. This is of much greater importance with shamas, European robins, and certain other softbills.

The normal introduction of potential mates is rather enjoyable to watch. The behavior is ritualized and species specific and, as such, is consistent in its progression. The entire introduction may take only an hour before the birds pair up.

Released into an aviary together the pair will become intensely curious about each other and close the distance between them. The male will begin to tilt and fan out his wings towards the hen, to show off his impressive wing pattern. He will squint his eyes in a peculiar way.

The hen may show a form of threat behavior by stretching her neck and head upward and standing stiffly on the perch. This standoff pose is seen between two males as well. She will move away if he approaches too quickly. A well-mannered male will then back off and wait.

Before long the hen will return and move in a bit closer. They will both move closer together and one of them will hop over the other to land on the other side. The male will display again intermittently.

A receptive hen will stretch upward, raise her tail to present her light colored under tail coverts and move it slowly up and down, while emitting a whining call. This sound is also heard when she invites him to mount her.

Soon one will try to touch the other's head or neck with their beak in a pre-preening gesture. The partner will present its raised

Figure 6.3-1 Male on right demonstrating intimidation or courtship display.

Figure 6.3-2 Standoff pose.

Figure 6.3-3 Jumping over each other behavior.

Figure 6.3-4 Hen in submissive and/or soliciting display.

Figure 6.3-5 Invitation to allopreening display.

Figure 6.3-6 Contact behavior of compatible birds.

neck and ruffled throat feathers. These are, incidentally, parts of their body they can't reach with their own beaks. If they accept each other, they will exchange cautious preening moves. If not, one or both will clap beaks as a threat and even fly at each other, but this is not common. Usually they follow each other and do more mutual preening, also called allopreening.

Allopreening is seen as a form of tension-reducing behavior. It diverts the escalation of aggression and bonds a flock. For example, this diversion is typical for white-eyes *Zosterops* spp. when they quarrel over food or perching spots.

It seems that once a pair roosts together for the night, it will develop strong bonds and stay mated for life, unless one is lost or removed. Breaking up the pair would be a consideration if one becomes non-reproductive and the other is still able and needed for breeding. Switching mates of breeding pairs creates more genetic diversity in the short term, but accelerates the inbreeding of the flock unless new founders can be obtained.

6.4 Preparation for Breeding

The breeding aviary must have shrub and tree plantings in the ground or in planters, with about one-third to one-half of it having plant material. It should reach close to the roof. The more cover the better, from the birds' perspectives, to build confidence and to nest and raise their brood. Where available, conifers can be cut and set in a pail with sand and water. If the sand is kept moist, the trees will remain green for two months or more, enough time for two clutches to be raised.

Live insect food is fed in increasing amounts as the breeding season approaches to provide extra animal protein to stimulate enlargement of reproductive organs and nesting. (Birds shrink their gonads and ovaries substantially after breeding season, in part to reduce body mass for migration). The food insects must be prepared to have the proper nutrient balance (6.9 and 6.10).

An often overlooked fact is that that minerals and vitamins, especially vitamin D and vitamin A, are critical and already in higher demand before the onset of nesting, for the hen to develop "healthy" eggs, embryos, and hatchlings. Without adequate vitamins and minerals packaged into the egg, we can expect infertile eggs, embryos dying in the shell, and chicks perishing at hatching time and within the first few days (chapter 8.5 and 8.6). From early on in the breeding season a dish with crushed eggshell and mineral supplement should be available for the birds, especially if the aviary does not have a natural soil substrate.

Poultry eggshell is collected in the kitchen. The lining membrane, which holds the shell together, should be peeled off to obtain close to pure minerals. If this is not done and the shell supply gets moist, the residual organic substances will decay and attract bacteria or fungus. Bake the eggshell in an oven for at least half an hour at 350°F (180°C) in order to kill any pathogens and to thoroughly dehydrate them for better storage. They can also be microwaved; however the thorough removal of moisture in a conventional oven is an advantage.

6.5 Breeding Territory

Each pair must have its own territory or breeding aviary. A good breeding aviary in this context is a space no less than 6 x 9 x 7 feet high (2 x 3 x 2.1 m).

If another pair is kept next to them, an opaque wall between the aviaries is essential or the pairs will fight each other and will not settle down to nest or rear their chicks. Even if the pairs have much larger aviaries, without solid partitions separating them the birds will spend much of their time squabbling through the wire partition.

There may also be a problem with bickering if two pairs face each other across a narrow walkway, forcing the erection of a sight barrier to end the disruption. I once put up a wood barrier but soon noticed that the neighboring pairs had discovered a knothole no larger than ¾ inch (18 mm), which allowed them to peck at each other. All nesting activities stopped until it was plugged. Cracks between boards must not be overlooked either.

It may be tempting to put another species into the breeding aviary, to "use all that space". This compromises success. Blue-breasted or Chinese painted quails, living on the floor and feeding at their own feeding station, should be compatible with pekin robins, and generally they are. The pekin robin pair will still nest with the quails as company, but I suggest removing the quails when the pekin robin chicks fledge. The chicks typically leave the nest all on the same day and many of them spend much time on the ground for the first day or two. The quails can become aggressive towards them, particularly if the quails also have a nest or chicks. Leaving the pekin robin chicks and the quails together just needlessly complicates the life of the pekin robins and their keeper.

Figure 6.5-1 Even a knothole can trigger territorial disputes.

Case Report

At one time a pair of pekin robins was housed with a pair and two extra males of Oriental white-eyes *Zosterops palpebrosus* in a well-planted aviary measuring 7 x 10 x 9 feet high (2.10 x 3 x 2.70 m). While the pekin robins carried nesting material at times, it never went beyond that stage. Finally the four white-eyes were

Figure 6.5-2 Several white-eyes flocking to the food dish and intimidating a pekin robin.

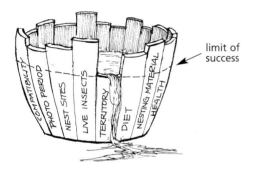

Figure 6.5-3 The "minimum law". The shortest stave determines the limit of success.

taken out of the pekin robins' territory and within a week the pekin robins began to build a nest, and subsequently raised a brood. So did the pair of white-eyes next to them, in a much smaller planted aviary.

There was no obvious conflict between the two species and one could conclude that the much smaller birds would have been under more stress than the pekin robins. Close observation of their interaction reveals an interesting conflict. The white-eyes have a different way of feeding at the dish and literally flock in unison to it and remain there until everyone has taken some food. In effect, they become one big bird and thus intimidate the pekin robins, which approach the dish one at a time. This competition strategy is remarkable and may have been the reason why the pekin robins felt unsure of their control over space and food.

All things must come together for success in breeding birds. The best way to illustrate this is by the "minimum law" demonstrated by a barrel with staves of different lengths; it can only be filled to the height of the shortest stave. In this case there was plenty of everything, except total ownership of the territory.

Case Report
I bred two pairs of pekin robins at different times in a large 20 x 13 x 14 foot high (6 x 4 x 4.2 m) outdoor aviary with tall, dense stands of bamboo and deciduous trees, that is shared with other species including a pair of blue-winged sivas *Minla cyanouroptera*, a pair of chaffinches *Fringilla coelebs*, a pair of European siskins *Carduelis spinus*, and a male European robin *Erithacus rubecula*. The blue-winged siva is closely related, and similar in size, to the pekin robin. One should expect considerable conflict between the two species, while the finches are less of a concern, but problems have been minimal, inasmuch as the sivas never nested.

Clearly the relationship of space and complexity within the aviary environment can overcome limitations.

In another indoor aviary 9 x 7 x 8 feet high (2.74 x 2.13 x 2.44 m) with an attached, larger outdoor enclosure, a pair of pekin robins raised a successful brood while sharing their space with pairs of coal tits *Parus ater* and blue tits *Parus caeruleus*.

A different pair was tried the following spring and while they built their nest without concern for the other birds, they repeatedly dismantled a chaffinch's nest. It may have been accidental in searching for spiders in the shreds of spider web the finch used to build the nest, or a territorial response. Eventually the pekin robins were removed to allow the chaffinch to nest successfully.

It should be noted here that, during breeding season, it is unwise to move pairs from aviary to aviary, or to switch mates in an attempt to improve nesting success. You may find that just when you think the pair is not accomplishing their task and you move them, the hen will lay her eggs on the aviary floor (without a nest), and troubled by stress-induced egg binding to boot.

It pays to be patient. Young pairs in particular progressively improve their behavior pattern during their first nesting attempts and may be more successful raising a brood by the end of their first breeding season.

Placing two pairs in the same breeding territory can have disastrous consequences since one pair may literally eliminate the other pair, killing them within a few hours.

Case Report
I had this tragic experience with two pairs of pekin robins that were transferred to my facility from a holding aviary where they had been kept, with four other pekin robins, in a small flight for the winter months.

The two pairs were compatible in my small plantless quarantine enclosure and were moved to a large planted aviary after two days. I had to be away on business for most of the day and on my return found one of the males near death. He was scalped, exposing the scull's connective tissue. There was considerable blood loss and the bird was in shock. Although hospitalized and treated intensely he expired in forty-eight hours. The extra hen was also traumatized and died of shock overnight. Necropsy showed that she had absolutely no food in her digestive tract; the harassing pair prevented her from feeding.

Two of the four birds apparently had paired during the flock situation and the moment they identified the new environment as a suitable breeding territory, they spontaneously attacked and effectively killed the other birds. This took place on March 15th about four weeks before I normally detect nesting behavior. Clearly the pair was already in the mood to establish a territory.

In my estimation, had there been three pairs and perhaps a number of other bird species present, this behavior would not have erupted. My Australian colleagues have seen this spontaneous, murderous behavior with splendid wrens *Malurus* spp. when two pairs are released into a potential breeding habitat, and confirm that this response would not be triggered with three pairs or more in that situation.[1]

The upshot of this is, don't put two pairs or an extra bird, especially a male, together in a suitable breeding environment in the approaching breeding season. The loser seldom survives the scalp-

Figure 6.5.4 Splendid wren in a territorial confrontation.

---

1. David Holmes et al., pers. com. 2004.

ing attacks. I was given a male that did recover in spite of it, but is left with a permanent bare head and scars around the eyes. (However he did contribute his genes in the following breeding season.) My false illusion that the birds would be inhibited from fighting with each other in new and unfamiliar surroundings was shattered with the unexpected and tragic loss of a promising breeding pair. Because of this experience I now try to have all breeding pairs sorted out and housed in their respective aviaries by December.

6.6 Nesting Material

Assuming it is spring, days are long, and live food has been fed, we will see hens and males carrying nesting material.

Long strands of raffia, about 8 inches (20 cm) long are favored to begin the structure of the nest rim. It is important to keep the length of the strands that short to prevent rare events where a strand of raffia gets wrapped around a bird's foot or other body parts. In one grotesque case I found a hen hanging by its neck from a branch, too late to be rescued. The same hen had been hung up a few days earlier by its foot. I had not experienced these problems before and suspect that the old bird, perhaps, lacked mobility.

Raffia is offered to help the birds anchor the outer framework while coconut palm fiber is provided for the lining of the nest. Some pairs have used fallen bamboo leaves for the lining, but dry grasses and rootlets are also used.

Cotton string is very enticing to nesting pekin robins, and more problematic. I once had a pair that dismantled a Chaffinch nest; the pieces of cotton string were of greatest interest to them for nest construction. However, it was that very type of material that strangled another hen in a friend's aviary.

I have observed about 100 nests being built. Generally they have been between 3 and 5 feet (90 to 150 cm) above ground, except in a walk-through aviary where the thickest bamboo leaf cover was 10 feet (3 m) above ground. Two horizontal twigs branching off to form a "V" create an inviting nest site.

The male seems to be the one who chooses the nest site by vocalizing with a peculiar rasping call and crouching at the selected spot. He will weave the first strands of raffia or grass to create the nest rim. The hen soon joins in and both work on the nest. The hen does more of the finishing work, and if a nest is reused while the previous fledged brood is still dependent on feeding by the parent, the hen does all of the refurbishing, since the male is occupied feeding the fledged chicks. Re-nesting before the chicks are weaned has unique problems, as we will explore later (6.7).

At times a nest is completed in one to two days and other times it takes many days, all depending on motivation. The nest is fairly

Figure 6.6-1 The first woven strands of raffia mark the spot to position the nest support.

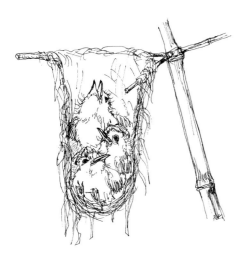

Figure 6.6-2 Without support a nest may fail as chicks grow heavier.

loosely built and its strength fully dependent on the anchorage of the longer strands of nesting material and the substructure to which it is attached. Some nests fail and the eggs or brood is lost. Nests without nest support can stretch with the weight of the growing chicks and form a narrow tube, which creates problems for the chicks to be properly fed since they become layered. I create nest supports to avoid this.

6.7 Nest Supports

A simple wire basket made of two to three crosspieces, connected to a rim, suffices to hold a nest in position and in shape. The hoop should have a diameter of 3½ to 4 inches (9 to 10 cm) and the depth of the basket is about 2½ inches (6 cm). The 16- to 18-gauge wire (about 1.5 mm in diameter) can be purchased with green plastic coating in garden supply stores. The wire basket is wrapped with raffia to resemble the started nest.

The nest support must be put into place shortly after the male starts to weave the first strands. If it is done much later, when the birds have become familiar with the nest's shape, they will become concerned that their nest has been discovered. The birds are best shifted into another aviary or the service corridor when the nest support is placed. Most pairs accept the helping hand, but on rare occasions the male will get annoyed and start a new nest in another site. Fussy pairs are better left alone.

Case Report

One male tried to start hopeless nest constructions on the outward reaches of bamboo branches and refused help on three beginnings, until the hen got anxious and simply added a few bits of vegetation into a supplied nest basket and laid a clutch of three eggs. The male did not want to take shifts to incubate and the nest was abandoned. (When a male does not participate in the original nest building, he rarely will help with the incubation, and sooner or later the hen gives up.)

In one atypical case both began to work on different nest sites. They could not decide to work on the same nest and wasted two precious months of the breeding season. They earned the name "the yuppie architects". Finally, in late August they got coordinated and managed to hatch two chicks. This behavior was repeated the following season, however three chicks were raised two weeks earlier, at the beginning of August.

It is suspected that the indecision and delay in the two above cases could be caused by the location of the aviary because it receives practically no sunlight. Pairs will be relocated to check this hypothesis.

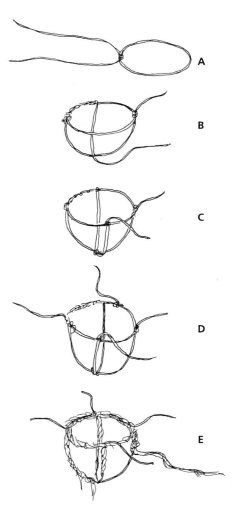

Figure 6.7-1 Sequence of creating a nest support.

A. A plastic-coated wire, approximately 30 inches (75 cm) long, forms the nest rim.

B. The long protruding ends create a cross to form the nest bowl, with two short ends extending beyond the rim for anchor ties.

C. & D. Another 10 to 12 inch (25 to 30 cm) wire is added to fix two more anchor ties in position.

E. The wire basket is wrapped with raffia to camouflage it.

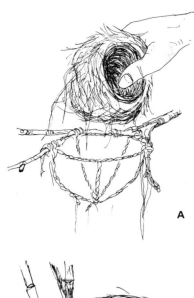

Figure 6.7-2

A. All the material can be removed from a nest support.
B. Only the intact inner coconut-fiber lining is taken out.
C. The liner can be used for an artificial nest in hand rearing birds.

Some pairs will start nesting with a nest support previously placed in the aviary. They seem to be attracted by the round shape of the rim. If there is a lack of good nesting sites or construction experience, a pair may struggle for days without making any progress. A nest support slipped in place usually ends the frustration for both the birds and the breeder.

The nest support has another real advantage, namely to delay re-nesting. The bulk of the nest material or just the compact inner coconut fiber lining can be easily lifted out of the nest support. (As an aside, the coconut fiber portion is very firm and can be used, after disinfecting it, as a hand-rearing nest by simply placing it into a dish.)

In an aviary, the birds will raise consecutive broods in the same nest. It is unlikely that this is generally the case in the wild, since the parents will travel throughout a large territory with their chicks for about two to three weeks after fledging. In an aviary situation, with plenty of food on location, the parents will start a new clutch of eggs within days of the chicks leaving the nest. The next clutch of eggs will then hatch before the older chicks are weaned. This creates conflict in feeding two sets of chicks.

Removing the lining from the nest support forces the birds to rebuild the nest, thereby helping to space the consecutive hatching of chicks (6.16). More time passes before the next clutch hatches, allowing time to adequately wean the previous clutch.

The survival rate of chicks in re-used nests also appears to be lower than in new or freshly lined nests. The accumulation of organic debris certainly creates a favorable environment for bacteria, fungus and parasites. While I have not collected enough data to prove this hypothesis, I have strong suspicions that the rate of failing nestlings is higher with re-used nests. Improved hygiene and higher motivation — if the nest is rebuilt or built anew — probably play a role in chick survival. Since the parents like to return to a successful site, the body of the nest can be lifted out of the nest support and destroyed or disinfected. Fresh nesting material is then provided to the birds to rebuild a cleaner nest interior.

6.8 Mating and Egg-laying

As a rule, mating does not occur until the nest is substantially or totally completed. The pair will copulate several times for two or more days. When the parents stop working on the nest, we can expect eggs within two to four days. Incubation begins once the second egg is laid. Most eggs are laid in the morning hours.

Sometimes you will see a hen with signs of stress; fluffed up and sitting still on a branch or even on the aviary floor. She is preparing herself for the strenuous task of producing an egg that weighs about three grams, or 11 to 13 percent of her body weight. It is best to leave her undisturbed. She will then go to the nest and lay the egg and soon be back to normal behavior. Other times she may be in the nest to lay, and the effort of producing an egg will be much less apparent.

If a hen becomes seriously egg-bound her health may deteriorate and the bird should be treated (chapter 11.2.6).

Figure 6.8-1 Typical posture for an egg-bound female not on a nest.

6.9 Incubation and Hatching

Most hens lay three eggs, some four, but very rarely five. A hen that lays four may do this for more than one clutch. Sometimes old hens only lay two eggs. The incubation period lasts for eleven to twelve days. In most cases the chicks hatch on the same day, even though — depending on clutch size — the last egg laid would only have been incubated for ten to eleven days. The hatching of the last egg may be delayed for about twenty-four hours, but seldom longer. A small weak chick may hatch, but often not survive to weaning age.

As a rule, the adults will remove any traces of the eggshells and dead chicks from the nest and carry them as far away as they can so as not to reveal the nest site. Sometimes eggshells are found that have been pushed to the outside through the mesh wire. Likewise, evicted chicks may be found at the perimeter of the aviary, which is a good place to start looking for missing nestlings. Eggs that do not hatch are often left in the nest while the chicks are also fed in the nest. This was also observed with silver-eared mesias.

It is widely advocated that one should not inspect the nest during incubation or the nesting period, with the necessary exception of banding the chicks. Even during nest building the parents are sensitive to being watched.[1] The moment you look straight at the birds, they will freeze at the nest and remain motionless until you have left again. You will see this behavior throughout the nesting period.

If the nest must be approached, then the parents should be shifted out of sight. Disturbance of the nest may give the parents the signal that is doomed to fail. The chicks are then abandoned or thrown out of the nest and a new nest is started in another location. The hen may also re-use the same nests. The tolerance of disturbance varies from bird to bird. In time there should be fewer problems by disturbances with domestically raised pekin robins, since super sensitive traits will not express themselves in the progeny, due to failing survival.

I made a very useful tool that is a small mirror glued to a pliable wire, to adjust its desired angle, and attached to a long stick. (I must have gotten the idea sitting in the dentist's chair.) It not

Figure 6.9-1 Freezing posture.

Figure 6.9-2 A mirror on a stick is a useful tool to check nests.

---

1. Mike Manley, pers. com. 1999.

only allows a peak into a nest well above eye level or in dense shrubbery, but more importantly, the birds don't react to the approach of the small mirror with much objection and, despite some scolding, will settle down much sooner than in a direct approach. I use this tool if I can't shift the birds away from the nesting area.

Observations indicate that the males are pivotal in the flow of events. Males clearly determine the nest site and start construction of the nest; they are the more vociferous when any disturbance occurs, and the ones that more often show signs of abandoning nest behavior, or curtailment of incubation and feeding, before the hens do. In my experience males determine more of the outcome than females. A recently acquired security camera will hopefully shed light on these chains of events.

Case Report
A pair less than a year old started nesting, built a solid nest and began incubation, but the male discontinued spelling off the incubating hen, and both abandoned the nest two days later. On the second try, both birds incubated faithfully and hatched their chicks. I was present when it took place. The male broke out in scolding sessions directed at the nestlings. The hen brooded them, but was put off the nest by the male. All signs indicated that he saw the chicks as alien creatures to be a threat to the nest. The hen returned to the nest for the night to brood them properly, but the chicks were not fed by either parent and, as a result were lost. On the third nesting both parents laid down a picture-perfect performance and raised two strong chicks in the same season. They had re-used the old nest on the second attempt and built a built new nest in a different location on the third.

Figure 6.9-3 A hatched chick can confuse a young male nesting for the first time.

6.10 Feeding Chicks

Providing the parents with proper food to pass to their young is the keystone of successful chick rearing. It starts with feeding the parents correctly to get them physiologically into good breeding condition: development of ovaries, gonads, calcium/vitamin level in the body of the hen, parasite control, immune system support and best possible health generally. Healthy birds equal healthy eggs, equal healthy embryos and hatchlings.

In my experience pekin robins and silver-eared mesias will only feed live food or very freshly killed live food to the chicks while in the nest. Frozen crickets and waxworms (chapter 9.2) have been ignored even if the chicks were actively begging for food. Interestingly, immobile, but "alive" insect pupae are taken without hesitation. This may be a "safe food response" since dead insects are quickly affected by undesirable bacteria, which would be harmful to chicks.

Crickets, grasshoppers, hairless caterpillars, termites, earwigs, smaller species of dragonflies, and especially spiders, are eagerly accepted and prepared for the chicks. Earthworms *Lubricus* sp. and slugs *Arion* spp. are not eaten by the adults or passed to the young. Maggots *Calliphora* spp. and *Musca* spp. have been ignored (even though they were raised on a cereal-milk-yeast formula) as have whiteworms *Enchytraeus albidus* and sow bugs *Oniscus* sp.

Figure 6.10-1 As the chicks mature the parents will bring several insects at once to the nest.

The parents feed their young throughout daylight hours. The daily portion of insects can't be offered at one time if the insects must be killed to prevent them from escaping. Small amounts should be given four to five times a day to ensure the insects are fresh. We can observe the birds carrying several mealworms or waxworms around in their bill waiting for the chicks to feed again. They will tempt the satisfied chicks repeatedly and eventually drop the insects to get fresh ones. This causes waste on one hand and shortage of food when the chicks are ready to feed again on the other; hence the food dishes need to be replenished periodically.

A more time saving way is to set up self-feeders for feeding live crickets and mealworms. These are plastic storage boxes about 10 x 16 x 11 inches high (25 x 40 x 28 cm). A forked stick can be placed on the bottom to train the birds to hop inside and catch their own crickets or mealworms. The crickets do not need to be killed in that case. If necessary, the self-feeder can be stocked up once a day. The food insects are given a lettuce leaf, slices of orange, and food formula to sustain them. Mineral/vitamin supplements can be added as well by "dusting" the crickets with it. The feeder is cleaned daily. This method allows a person who has to leave the home during the day to still raise pekin robins.

Figure 6.10-2 A plastic storage box makes an excellent live cricket feeder.

A word of caution:
During one summer with an extraordinarily high wasp *Vespula* spp. population, the self-feeders were plundered by hordes of wasps. They not only competed for food insects, they also intimidated the parents entering the feeding boxes, and subsequently two nests with young failed during the peak of the wasp invasion. This occurred despite trapping hundreds of wasps and frequently moving the boxes to different locations. This points out the importance of periodically checking on the parent chick-feeding patterns.

Waxworms are more difficult to offer because they simply crawl away and can't be contained, even in open, smooth, glass containers. They are fed out of hand or killed and fed immediately. This may be just as well since the caterpillars have very tough and sharp mandibles to enable them to chew through soft plastic

Figure 6.10-3 Parent birds will make small hops on the nest rim to entice the chicks to gape.

and wood. While the parents are very diligent in killing the larvae, it is comforting to know that the waxworms are dead one way or the other, before being swallowed by very young chicks.

The amount of live food that is consumed grows with the chicks and reaches a peak before fledging. A family of five may eat 300 and more insects per day. To ensure sufficient supply, and quality insects, it may become necessary to maintain your own insect cultures. Crickets are rather expensive to buy; yet thousands can be raised for very little money (Cultivation of Live Food, chapter 9).

The parents signal their arrival with food by tapping the nest rim with small hops to which the chicks respond by stretching their neck and heads up, gaping for food. The parent may also emit low-level rasping calls to encourage the chicks to gape. As the chicks begin to open their eyes, on approximately the sixth day after hatching, they start to respond to their approaching parents by vision.

After feeding the adults wait at the nest to receive the fecal sacks from the chicks. The fecal sacks contain excrement encased in a mucous membrane. At the onset they will consume the fecal sacks and later they carry these away to the perimeter and even push them through the mesh wire. This is an innate behavior to guard the nest against detection by predators. The nest is kept very clean and any foreign objects are removed. This can be of concern when the chicks are banded too early (6.13). The parents will remove the feces until about two days before fledging. By that time the chicks will begin to defecate over the rim of the nest.

6.11 Early Loss of Chicks

There are frequent reports on loss of chicks during the early stage of development when they are about four to five days old. The parents may throw the chicks out even if they have not died. At times one can see lesions on the head, neck, and back indicating that the parents actually attacked or killed the chicks. The reason for this is not fully understood. In my observations it relates to the failing health and vitality of the chicks, or major disturbances. The lack of live food has also been suspected in some cases; however that was not a factor in the birds I have worked with.

Lack of vitality may already have its origin in low levels of mineral and vitamins supplied to the hen with the result that subsequent "undernourished" eggs have inadequate reserves to get the chicks through their first few days of life (chapter 8.5 and 8.6).

Normal feeding behavior is an inter-play of stimulus and response. Essentially, the hungry chick gapes and the adult will feed it. If the chick is satisfied, it puts its head down and crouches in the nest. Older chicks will close their eyes and turn their heads away to signal that they are no longer hungry. If it is close to being full it will not swallow a food item vigorously, so the adult will pull the food back out and give it to another chick, or drop it to get fresh food.

This pattern is disrupted when a chick is not well, for example, due to some form of infection or nutrient deficiency. If it is too weak to swallow with the right response, it is not fed properly and loses more energy and vitality, and a downward spiral sets in.

Some aviculturists speak of chicks or the parents "getting tired" of the same food, and have observed that changing to other insects will restart an active feeding response. I have noted that offering spiders in particular, but also cave crickets *Ceuthophilus* sp. and other wild-caught insects, stimulates declining feeding behavior.

Presumably, the birds instinctively search for variety to present a wide and full spectrum of nutrients, minerals, and vitamins to the growing chicks. The sight of a spider or different insect raises excitement and stronger feeding impulse to overcome "loss of interest". Once the chicks ingest sufficient food, their vigor may return to combat any slight ailment or deficiency.

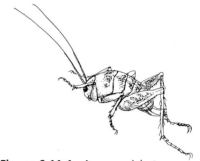

Figure 6.11-1 A cave cricket.

Another opinion is that the chicks die after a few days if they are fed waxworms and mealworms. This should be anticipated if the birds are fed with insect larvae that are raised solely on wheat bran, the standard medium for most commercially raised mealworms. Furthermore mealworms bought from many pet shops are often not fed, but held in a refrigerator. The combination of little food content in the larvae and the lack of available calcium in wheat bran create inferior food with a problematic calcium-phosphorus (Ca:P) imbalance (chapter 8.6).

Mineral deficiency will reduce the vitality of chicks and that alone may be part of the cause for removing chicks from the nest or the chicks failing to survive. Food insects must be prepared or "gut-loaded" prior to feeding. The best way to achieve this is raising and preparing the food insects on location (chapter 9).

The ratio of Ca to P should be 2:1 and no less than 1:1. The diet should roughly contain about one percent calcium, which can't be supplied via inferior insect larvae. Special attention must be paid to vitamin D provisions, because it is critical for calcium absorption. This is particularly important if neither the parents nor the chicks have exposure to direct sunlight to convert pro vitamins. Rickets, indicated by leg and bill deformation ("rubber beaks"), and an inability to fly after fledging, are typical symptoms. There are also less obvious deficiencies body functions affecting metabolism, nervous system, etc.

More on this is discussed in the chapters on Feeding (8.5 to 8.6) and Health Care (11.2.8).

Additional calcium must be added to the insect diet in one form or another. Chicks that are raised on commercially produced

Figure 6.11-2 Injecting supplements into a waxworm.

Figure 6.11-3 Adult "hammering" a food insect against a perch to prepare it for feeding.

Figure 6.11-4 Overheated chicks will stretch their necks over the nest rim.

crickets are less vulnerable, since crickets have higher calcium content than mealworms or waxworms. A stopgap measure is to inject a dissolved vitamin/mineral concentrate into waxworms and mealworms or other insects. One can also coat the mealworms very lightly with margarine and dust the vitamin/mineral supplement powder onto them; it ensures that some of it is delivered to the chicks. Halibut oil can also be used; the mineral powder will adhere to the halibut oil and provide vitamin D_3 and vitamin A at the same time. There are also liquid supplements on the market that can be used to lightly coat the food insects.

Simply dusting the waxworms or mealworms is ineffective, since the adults will "hammer" the larvae against the perch to kill and soften them for feeding. The powder is then knocked off and wasted. I found that adult birds might be selective and reject certain supplements, even if they are injected into waxworms. One must then to switch to different products.

The feeding of high quality insects, nest hygiene, and controlled disturbance of the nest site emerge as key factors to prevent chick losses. The earliest and latest clutches in season are most vulnerable to chick loss. Short day length and lower ambient temperatures are likely responsible, and one should not be discouraged if chicks fail to survive in the shoulder season (6.17). I suspect that the poorly insulated nests of pekin robins create a cold zone under the chicks, which interferes with food digestion and metabolism.

Environmental conditions must be monitored. The ambient temperature can fluctuate significantly in an outdoor aviary. Sun-rays behind glass will raise temperatures to dangerous levels. The chicks will show stress from over heating by stretching their heads over the rim of the nest. Shading must be added for the nest site to curb this problem.

Predator control is another aspect to be aware of to prevent losses of nestlings (chapter 4.3).

6.12 Fledging

Even if the brood has been raised for ten to twelve days to fledging age, there are often losses again after three to four days out of the nest. A healthy, strong chick can be seen perching up high in the aviary and able to fly from branch to branch on the first or second day after fledging. The chicks are however not of equal strength, partly because they hatched from eggs laid over three to four days, which makes it difficult for them to catch up to their older siblings (6.8).

The chicks' normal behavior and strong urge is to huddle together with siblings in the cover of a bush or a place they feel secure. In a sparsely planted aviary they seek higher perches. This

Figure 6.12-1 Posture of healthy chicks after fledging.

is important to recognize, because the weaker chick will not be able to fly up, and must hop from branch to branch to get close to its siblings. It will try time after time and make endless attempts, only to become exhausted, most anxious, and severely stressed. This reduces its energy to the point where we may witness the parents' impatience when feeding it by withdrawing the food item before the chick has swallowed it, and soon it loses ground beyond recovery. Resting and sleeping with its head hanging down is a typical sign of trouble. It may become necessary to transfer the bird to a hospital cage for hand rearing.

Pekin robins seem to fledge before they are able to fly well, much like the American robin *Turdus migratorius*. Their survival depends on good hiding places, where the parents can feed them. Gradually the birds progress and master their power of flight. The presence of low bushes and cover in the aviary is, therefore, necessary to provide hiding places for the chicks.

Stress-induced muscle burnout (chapter 11.4) can be so severe as to cause paralysis due to muscle damage and shock. This failure stems from the lack of a suitable environment, a situation we can do something about. Unless the chick has a quiet, warm place and receives food, it will succumb.

Figure 6.12-2 Fledged chicks are determined to roost together.

Case Report

To illustrate the problem of muscle burnout, I once raised a chick by hand. We had to travel by car on the day of fledging, and the only way to provide food throughout the day was to take the bird along in a small cage on the back seat of the car. The bird was completely tame. It fed well and appeared unusually active and precocial, flying back and forth in the traveling cage for most of the day. Due to its exhibited strength it was called "super chick". The next morning it sat on the floor of the cage fluffed up, eyes half closed and unable to perch, let alone fly up. It showed paralysis in its legs. It was clear that it had not contracted an infectious disease that would show such dramatic symptoms in that short period of time.

The chick was kept warm and force-fed soft food, with added electrolytes and vitamin mineral supplements, for a lack of better knowledge about how to treat it. It became so ill I expected to lose it imminently, but it gradually improved. It took a week to regain the ability to perch, then to fly, and to fully recover.

Figure 6.12-3 A chick with beginning signs of exertion myopathy or loss of general condition.

Figure 6.12-4 Paralysis of legs caused by exertion myopathy.

This was a classic case of muscle burnout and resulting shock. It was also an eye opener to understand that a healthy chick at fledging could deteriorate so rapidly after desperately trying to find a hiding place. The endless efforts to jump up on a smooth wall, and repeated falls because the perches are out of reach, simply tires out a small chick, which sits in a nest until the day of fledging. Such enormous physiological and psychological stress must be avoided. If need be, more plant material can be brought into the aviary. It will upset the entire family for the moment, but once they settle down it may well save the chicks.

The parents will continue to exclusively feed insects until the chicks are eighteen to twenty days old. The first non-insect food the parents offer to the chicks is a home baked egg cake (chapter 8.11). Once this stage is reached, successful rearing is the norm. The egg cake has all the required food constituents, to which is added a vitamin/mineral supplement in powder form to end worries about dietary imbalances or deficiencies. Fruit is fed to the chicks following the acceptance of the egg cake.

The chicks begin to pick up food on their own after four to four and a half weeks. (This roughly coincides with the first visible yellow and orange feathers in the throat area.) The adults are so engaged and eager to feed their young, it delays independent feeding by up to ten days compared to hand-raised chicks (chapter 7.5).

6.13 Weaning

Weaning in mammals means curtailing the suckling of the mothers milk or, conversely, bottle-feeding of the young animal. In the avian context it means that the young birds become independent from feeding by their parents or hand feeding by a foster parent.

As the chicks mature they must be removed from their parent's aviary. They will interfere with the raising of subsequent broods and cause conflict between the young and adults, particularly between the male offspring and the sire. The availability of an adjoining aviary is ideal for the weaning process. The family is given access and food to both spaces until the young are simply blocked from returning to the breeding aviary. Visual and preferably audible contact between parents and chicks should be eliminated.

This is often high time since the hen tends to restore the nest and begin to lay again long before the chicks are weaned. In some cases the hen lays within four days after the chicks leave the nest. The male does most of the feeding and the hen most of the incubating when a fledged clutch is cared for at the same time.

The eggs will hatch in eleven days at a time when the fledged young are less than thirty days old and a week away from the normal weaning time. The chicks sometimes try to roost on the nest rim with an incubating parent and even settle down in the nest, if the parents miss their change of guard. This will interrupt incubation, but usually the adult manages to edge the chicks away.

Greater trouble starts when the eggs hatch and the old chicks follow the adults to the nest to take the food intended for the

Figure 6.13-1 Older sibling competing for food.

newly hatched. By that time the older siblings must be weaned. This can be achieved with the help of egg cake and live insects, plus careful surveillance, to ensure that the birds consume enough food. Removal of the nest body from the nest support buys the breeder a few extra days to spread the events so that weaning is better managed (6.7 and 6.16).

6.14 Banding

Banding, or ringing, means the application of a metal or plastic band around the leg bone (tarsus) to identify a bird. A closed band or ring has an inside diameter that will only allow it to be slipped over the foot joint only at a very early age, a few days before the birds fledge. This is a form of proof that the bird has been raised under human care. Split plastic bands in various colors help identification at a distance.

Closed, coded, registered leg bands or rings are required for any export and import of CITES-listed birds, but also as proof within a country for legal acquisition and possession. The closed metal band, usually aluminum, is of a particular inside diameter for a given species. It is regulated in various countries.

For pekin robins and mesias, I use the size K with an inside diameter of 5/32-inch (3.94 mm). In Europe 3.5 mm (.14 inches) is the issued size for pekin robins. The Avicultural Advancement Council of Canada (AACC) has adopted size K bands for pekin robins and silver-eared mesias. The next size below that is J, .13 inches (3.30 mm). The latter is too small to be applied once the chick is six to seven days old. It would fit sooner after hatching. This is restricted, however, by the earlier-mentioned behavior in which the parents keep the nest free of any fecal and foreign matter. The chicks will not have enough pinfeathers, down cover and body size to prevent the parents from detecting the leg bands. If they do, they will attempt to remove it from the nest, chick attached.

I target day seven with "K" bands for pekin robins and day six for silver-eared mesias. Larger bands, or banding too, lead to complications by trapping the hind toe or slipping over the shank/tarsus joint.

A special portable tray is used, which can be hooked on the wire mesh. It has a dish with clean tissue paper to receive the chicks and a shelf to have the bands ready, plus a spot for a dab of petroleum jelly.

If at all possible the parents should be shifted out of the aviary because their frantic scolding can trigger the chicks to panic and jump from the nest up to three days before they would normally fledge. Returning chicks older than eight to nine days to the nest is usually unsuccessful in this situation. It is not critical to do the

Figure 6.14-1 An over-sized closed band can slip over the tarsus joint.

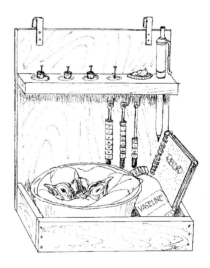

Figure 6.14-2 A banding tray, with closed bands ready for application, Vaseline, additional bands, tissue-lined dish, pen, and note pad, that can be hung on the wire mesh walls, organizes the process for swift banding.

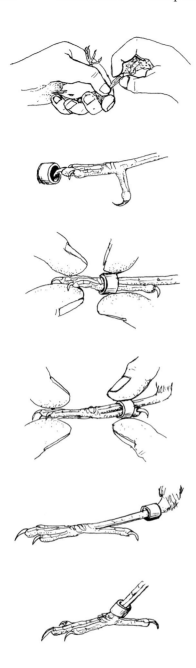

Figure 6.14-3 Sequence for applying closed leg bands. The three front toes are stroked together with the application of Vaseline; the band is slipped over the "bundled" front toes; the toe tips are held firm and the band is moved over the foot joint with gentle rotation of necessary; the band is pushed all the way up the tarsus until the hind toe comes free.

Figure 6.14-4 A hind toenail trapped in the band must be freed as soon as possible.

Figure 6.14-5 Banding spoon for split plastic leg bands.

banding swiftly if the parents are locked away and neither is it critical to do it near the nest. In fact, if the parents are at the nest, the chicks should be taken into another room so the parents settle down at least until the chicks are put back in the nest.

All the chicks are taken out and placed in the bowl, and then one by one banded and returned to the nest. To band them, the three front toes are dabbed with a trace of petroleum jelly and stroked together to form a bundle, the band is pushed over it and the foot joint onto the leg with the hind toe pointing backwards and aligning with the tarsus. The band is moved up until the hind toe comes free. If it does not, a toothpick or small nail is inserted to bring it out.

It is important to firmly hold the front toes while the band is pulled and not pushed over the foot joint and onto the tarsus so not to compress (telescope) the bone or kink and injure it. A gentle rotation of the band while it is worked over the foot joint may be necessary to pass it over the joint. The band numbers are recorded and notes are made of any features such as relative size of the chick. The chicks should be checked for several days after banding since the banded birds, especially smaller siblings with lighter bone structure, may accidentally catch their hind toe nail in the band. The bird must then be captured to free the hind toe.

It is common practice to band males on the right leg and females on the left. This can't be done with pekin robin chicks, since sexes can't be told at this early age. To make identification at a distance, colored, numbered plastic split bands can be added later when sexes are confirmed. These are applied with a special spoon-like tool, which is supplied by the ring registrar who supplies the bands. Split bands can be removed by prying the end apart.

There are also single-use aluminum bands that can be applied but not opened and closed again without breaking apart. Federal quarantine agents use these for identification and proof of legal importation.

6.15 Record Keeping

Record keeping is necessary to maintain a breeding program and it has become a requirement if you keep species with import restrictions. There will always be birds where banding with closed bands cannot be done, which leaves no hard evidence that they were raised in an aviary. Colored and numbered split bands are necessary, if the birds have not been close-banded. The authorities governing the pertinent legislation will generally accept well-maintained records for legal possession. A manual card system or ledger can be set up to track all birds.

With today's computers, record keeping can be done electronically. There should always be a back-up file with another computer, or a copy, periodically updated, made on CDs, stored in a different location in case one is destroyed by fire or other disaster.

Zoo organizations and aviculturists have established studbooks for many species and new ones are added from time to time. A studbook, currently maintained by the author, was established in the year 2000 to track individuals of the genus *Leiothrix* in Canada. This is a fundamental tool to maintain a meaningful conservation-breeding program for small population management to improve genetic diversity, and to avoid inbreeding. The studbook keeper assists in the coordination of breeding initiatives by the consortium of breeders (see also chapter 10.3).

Diaries are invaluable records, and essential to back up the data entry in summary reports and studbook records. Behavior observations can be compared from year to year to establish trends and traits within the flock to predict recurring events. Medical treatments, hatching dates, inventory changes, weather, diet changes, and notes on the food insect cultures are just some examples of what should be recorded. I found the best way to ensure that consistent entries are made, is to place the diary on the night table and make the entries at the end of the day.

6.16 Breeding Events Calendar

A review of numerous breeding seasons by several pairs reveals that a new clutch of eggs is laid very close to thirty days apart in an undisturbed environment with a proper food supply. The time from the start of the incubation to weaning is however around forty days, because of the overlap in producing two consecutive clutches.

The overlap of the hatching of a following clutch while the first is still fed by the parents becomes evident in the schedule below. The hen may already start her next clutch two to three days after the chicks have fledged. By removing the nest lining, extra days pass before the nest is restored and eggs are laid again (6.7). The delay better synchronizes the weaning and hatching of consecutive clutches (6.13).

Table 6.16-1
Sequence of consecutive hatchings by pairs re-using same nest (based on average records)

Day	First Clutch Event	Age of First Chicks (days)	Second Clutch Event
1	First egg in nest		
2	Second egg – incubation starts		
3	Third egg		
4	Possible fourth egg		
13	Eggs hatch (12 days incubation)	1	
24–25	Fledging (12 days of nesting)	11–12	
25–30	Chicks stay in seclusion	12–18	First egg
30–32	Chicks start moving around	18–20	Incubation started
34–35	Adults start feeding non-insect food	22–23	Incubating
38–41	Chicks begin to feed on own	26–29	Hatching
41–43	Chicks become independent	29–31	Second clutch chicks 1–3 days old
43–48	First clutch weaned (5 weeks old)	31–35	Second clutch chicks 3–5 days old

6.17 Breeding Statistics

Table 6.17-1 reflects seven-year breeding statistics on pekin robins *Leiothrix lutea* and silver-eared Mesia *Leiothrix argentauris*. Records were collected from a breeding program of the genus *Leiothrix*.

The program started with one pair of pekin robins in 1999. By 2005, a total of 11.5 wild caught founders, 1.11 F^1 and 1.0 F^2 offspring formed nineteen pair combinations. Eighty-five pekin robins were raised to independence (weaned) in the seven-year period. Regional studbook records are maintained. Most pairs that were together between April and September had three successful nesting attempts, and some four attempts per season.

Three pairs of silver-eared mesias were bred for one year each in 2000, 2004 and 2005. The data is presented for interest. The small sample does not allow one to set bench marks; however, lower success can be expected since mesias exhibited less tolerance to banding chicks in the nest and other disturbances. Two chicks were abandoned and one tossed from the nest after banding. Two fledged chicks, startled by a hawk, died of impact trauma after hitting a glass window. (Mounting dense branches of conifers in front of glass panes later eliminated problems with impact trauma.)

Only one of eighteen pairs of pekin robins exhibited a tendency to abandon their nest due to disturbance after incubation commenced. None abandoned chicks or removed chicks from the nest due to banding.

Table 6.17-1
Breeding statistics on pekin robins *Leiothrix lutea* and silver-eared mesia *Leiothrix argentauris* between 1999–2005.

Pekin robins	April	May	June	July	Aug	Sept	TOTAL
Number of nests	4	10	14	23	12	6	69
Average number of eggs laid	12	35	50	73	40	17	227
Number of eggs per nest	3	3.5	3.8	3.2	3.3	2.8	3.3
Total number of chicks hatched	7	28	41	55	29	11	172
% of eggs hatched	58.3	80	82	75.3	76.3	64.7	75.8
Total number of chicks fledged	1	13	22	47	19	9	111
Total number of chicks weaned	1	8	18	37	16	5	85
% of success vs. number of eggs	8.3	22.8	36	50.7	42	29.4	37.4

Silver-eared mesias	April	May	June	July	Aug	Sept	TOTAL
Number of nests	2	2	3	3	3	1	14
Average number of eggs laid	4	6	10	9	9	3	41
Number of eggs per nest	2	3	3.3	3	3	3	2.9
Total number of chicks hatched	1	4	6	6	8	2	27
% of eggs hatched	25	66.6	60	66.6	88.8	66.6	65.8
Total number of chicks fledged	0	0	3	4	4	2	13
Total number of chicks weaned	0	0	1	0	3	2	6
% of success vs. number of eggs	0	0	10	0	33.3	66.6	14.6

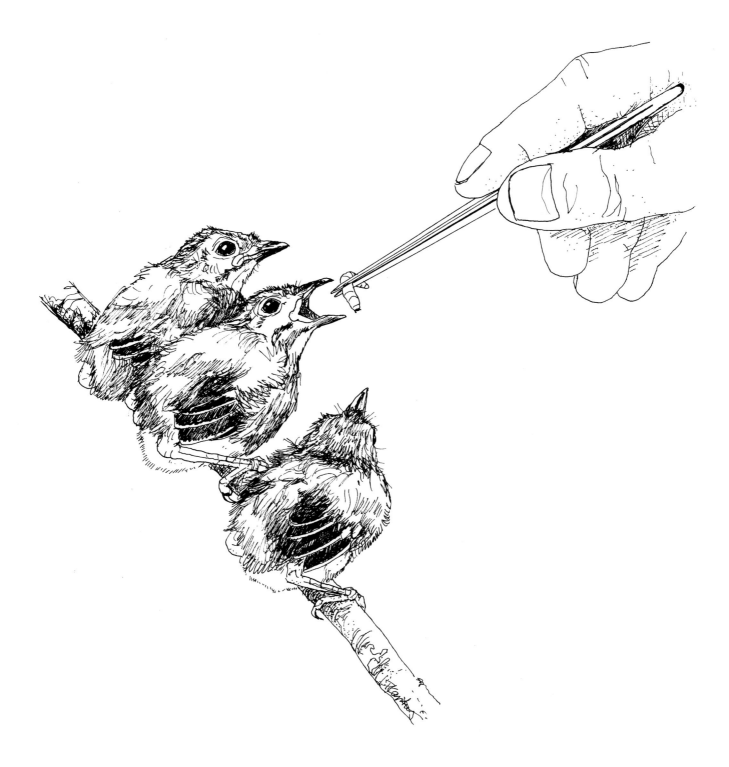

Figure 7.0-1 Hand feeding three newly fledged pekin robin chicks.

Hand-rearing

7.1 Introduction

Hand-rearing means feeding and caring for an animal by human foster parents until it is self-feeding. In aviculture it entails feeding, removing feces, maintaining a nest, and brooding chicks until the birds have become independent of artificial care (weaned).

In a perfect world hand-rearing is not necessary, but sooner or later the breeder is faced with the task of saving an abandoned young bird. Good husbandry and proper environment reduce the likelihood of such interventions. Fertile pekin robin eggs can also be incubated artificially and the chicks hand-reared, as noted below.

I am not in favor of deliberately removing and hand-rearing chicks to cause the hen to relay quickly to produce a high number of offspring per season. Pekin robins and other species are valuable birds for conservation breeding and anything that jeopardizes longevity, natural behavior development, or general health must be questioned. It may have application as a last-ditch effort in a species recovery program, to increase a critically small population.

The best chance for success in hand-rearing a chick is after its fourth or fifth day of hatching. Presumably, by then the parents have passed on the needed intestinal flora and immune system agents to the chicks, so they can cope better under the care of human foster parents. The chicks develop at an enormous rate and quickly become more robust and coordinated to swallow presented food, and especially liquids, without accidentally aspirating them. Birds, unlike us, do not have an epiglottis, the little flap that shuts off the larynx (windpipe). They close the glottis by constricting muscles. Newly hatched chicks are more prone to the dreaded and fatal fluid aspiration (chapters 7.4 and 11.2.1). Bacterial infection and diarrhea are other real threats to small chicks and good sanitation/hygiene must be practiced to reduce these risks.

There are various publications on hand-rearing pet birds, especially psittacines, which give detailed instructions on raising chicks and information about artificial formula. For four-day old and older pekin robins and mesia chicks, I found that the continuation of food insects offered by the parents, namely cricket bellies, waxworms, and mealworms, plus occasional wild-caught insects, is quite satisfactory.

7.2 Basic Equipment

The most basic equipment includes an artificial nest, a heat source, a thermometer, a soft cloth to cover naked chicks, and instruments to deliver the food items and to remove feces.

For a small aviary operation one can improvise basic equipment to hand-rear chicks. Only basic home-built rearing equipment is described here.

The chicks are placed in a nest, which is seated into a heavy flowerpot to prevent it from tipping over. A small plastic margarine container can also be used. We can use nest linings that have been removed form old but sterilized pekin nests (chapter 6.7). In any case, a nest must be lined with clean hay, coconut palm fiber, or other fibers that allow the chicks to exercise the grip of their feet for strong bone and muscle development. Soft tissue paper, cotton, and similar materials are not recommended.

Figure 7.2-1 Hand-rearing station.

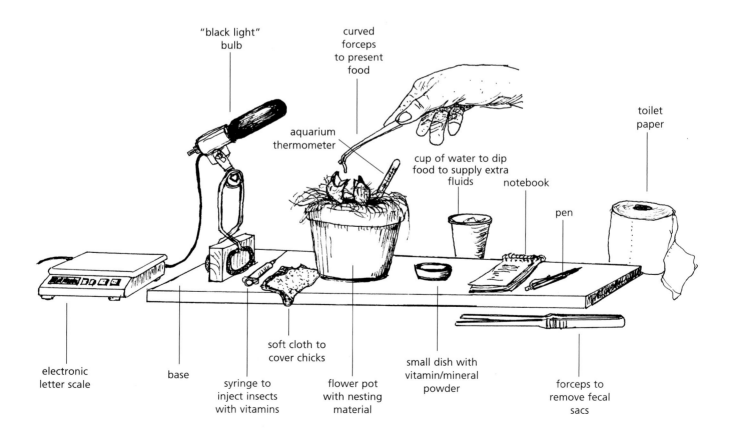

The chicks are covered with a piece of soft cloth while they are without down and pin feathers. Once the feathers begin to cover the chicks, this is no longer necessary.

Heat is provided by an infrared lamp (50 watt) or light bulb that directs its rays towards the body of the chicks. I prefer to use a "black light" bulb sold for reptile terrariums to give the chicks darkened night hours. Moving the lamp closer or further away regulates the temperature in the nest. This is a primitive way to regulate ambient temperature, which only works if the room temperature remains constant. Sudden sunrays striking the nest behind a glass window would be devastating. There must always be a thermometer in the nest to constantly monitor the ambient temperature around the chicks. An aquarium thermometer can be pushed into the nest lining next to the chicks. It can be read even if they are covered with a cloth. If the chicks huddle together and feel cold and clammy they are too cold and if they pant or try to hang their necks over the nest rim or out from under the covering cloth, they are too hot. Drafts must always be avoided.

Figure 7.2-2 Sunrays striking a chick behind glass will quickly overheat it.

I found that chicks older than three days do not need a nest temperature higher than 87°F (32°C). Their body temperature is, of course, higher. Most birds have body temperatures ranging from 104 to 107°F (40.5 to 41.6°C). The chicks metabolize their food and generate much of their own body heat, which they trap more efficiently once the down and contour feathers cover their skin. The temperature can be dropped gradually to room temperature by fledging time at three weeks of age.

It is more likely chicks will be lost with the temperature going awry on the high side than the low side. The nestlings bridge short periods of low ambient temperature. In nature they are left alone at times when a predator drives off the brooding parents. In fact, you will be terrified watching the adults in your aviary taking long breaks by having an extended bath and engaging in mutual preening (allopreening) sessions, before one finally decides to return to the nest of eggs or chicks, even in cool weather.

On one occasion I mounted a security camera at a pekin robin nest site and upset the hen so that she refused to return to the nest. The end of the day was approaching and, since normally only the hen incubates at nights, I thought that this clutch was in big trouble. It is not typical for a male to incubate at nights, leaving the eggs to chill beyond recovery. After three and a half hours I took the three cold eggs, candled them, and placed the only egg that showed development under another very old hen that had just failed to hatch her single unfertile egg. A lucky timing break: the old hen accepted and hatched the egg with her mate and brought up a nice chick. The eggs could have easily been chilled to the point of no recovery.

This episode, which demonstrates that the embryo in the egg can survive very long periods without being brooded, will help me to relax when in the future the birds take extended breaks from parental duties at nesting time.

7.3 Brooder Construction

For the few occasions it is needed, a simple brooder can be improvised at minimal cost. Such a unit is a good improvement over the basic equipment as described above, because of the thermostatically regulated ambient temperature in the unit. Essentially it is a heated container with controls to regulate temperature and a way to adjust humidity. Light bulbs can be used for a heat source. A plastic storage container with a drawer is convenient to access the nest inside. The size of water pan determines the humidity.

More advanced and technically reliable commercial brooders are on the market for operations with more extensive needs.

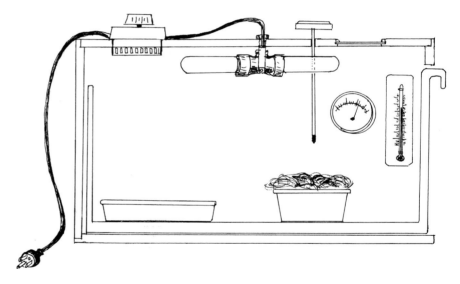

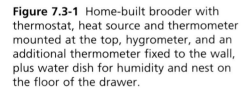

Figure 7.3-1 Home-built brooder with thermostat, heat source and thermometer mounted at the top, hygrometer, and an additional thermometer fixed to the wall, plus water dish for humidity and nest on the floor of the drawer.

Figure 7.4-1 Gentle pressure at the base of the beak facilitates gaping.

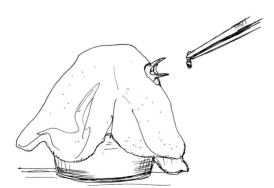

Figure 7.4-2 Chick covered by a cloth with its beak protruding through a small hole.

7.4 Hand Feeding Chicks

The chick may beg (gape) for food, but refuse to be fed for the first two to three meals and must be coerced to accept food by gently opening and placing food between the mandibles. Food is carefully pushed towards the back of the tongue until it is swallowed. Gaping may also be triggered by tapping the beak and/or whistling. For chicks which are intimidated by their keepers, a piece of cloth, used as a sight barrier, can be placed over the head of the chick with only the beak protruding.[1]

1. Chris Sheppard, pers. com. 2006.

To open the beak, the chick is held in one hand with thumb and index finger giving light and sustained pressure at the base of the beak until it opens, while the other hand is ready to place a food morsel into it.

Often the chick, particularly those older or fledged, will fight to keep the beak closed, making it necessary to gently pry it open with a toothpick or wooden tool created for this. You may think this is never going to work and feel sorry for manipulating the bird, but it rarely takes more than two or three force-feedings before the chick will gape and take food from you. Dull, curved sets of forceps work nicely to deliver the food items. Freshly shed, "white" mealworms or killed waxworms are the easiest to use for the first feedings. Later, more fragile abdomens of crickets can be fed.

Some people recommend giving the chick a drop of water. This is very risky, as is offering any liquid food. A frightened, struggling chick does not "drink" properly and instead may aspirate the liquid into the lungs. Avoid flooding the mouth cavity during feeding procedures. Fluid aspiration is usually fatal within a few hours.

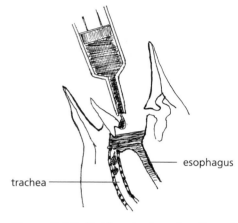

Figure 7.4-3 Fluid aspiration, flooding of the mouth leading to fluid accidentally entering the trachea.

Fluid aspiration is more likely to occur at the first feedings and with very young chicks. It is much safer and generally sufficient to dip the food into water before feeding it. Another way is to use a watercolor brush and transfer water (with electrolytes added) to the side of the beak.

The chick responds to the fluid adhering to the beak and deliberately "drinks it in", rather than trying to cope with fluid running down its throat. The key is that the chick initiates its own response to drink. Vitamins, electrolytes, and minerals can be given to the chick by injecting them into the worms or dipping the food into it. Electrolytes are important to support a weakened and dehydrated chick.

Figure 7.4-4 A watercolor brush is used to offer water.

Unless a chick has contracted some form of infection, most often in the digestive system resulting in diarrhea and related respiratory trouble, the chick should progress well and be feeding itself when about three weeks old (chapter 4.5).

Once the chicks have accepted their foster parent, they will stretch up and gape for food. Small chicks will do this when the covering piece of cloth is lifted; a light tapping on the nest rim will trigger gaping as well. We can also observe the parent birds "tapping" by doing small hops on the nest rim when they arrive with food.

If you have rescued two or more chicks and left them quietly for some time, you can trick them into gaping by tapping the nest rim and quickly offering food. The chicks will be spurred on by feeding competition, and this will save you manipulating them at the onset, as described above. Once the chicks have enough food they will no longer gape. You may face a situation where a chick is

Figure 7.4-5 Tapping of the nest rim triggers gaping.

Figure 7.4-6 A fecal sac is caught with forceps.

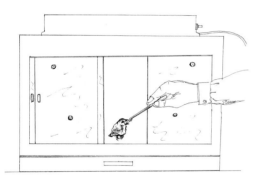

Figure 7.5-1 Feeding a fledged chick in a hospital cage. The three sliding front panels provide easy access within a secure space.

Figure 7.5-2 A chick presents its neck for preening.

already weakened to the point of not reacting or swallowing well. You will have to make a judgment call as to how much and what kind of food to provide. A note of caution: overfeeding must be avoided.

Young chicks, from four to six days old, are fed at hourly intervals over a sixteen-to-eighteen hour stretch in a day. Gradually the intervals can be lengthened to every three hours over twelve to fourteen hours per day by fledging age. It is, however, more natural to feed more often. Smaller, frequent portions are better digested than few large "stuffing" meals, because pekin robins do not have a crop to store large volumes.

Once a chick is fed, it often immediately presents a fecal sack by thrusting its posterior towards the edge of the nest. This is quickly caught with forceps to keep the nest and chicks clean. The forceps for this are not the same as the forceps used for feeding. Defecation and food intake are recorded in a notepad, which should be kept for future reference.

7.5 Fledging Hand-reared Chicks

As the chicks approach fledging they will start to defecate over the nest rim. Keeping the nest clean may require the replacement of the fiber lining. Beware of the approaching fledging day. It may be at an age of ten days or sooner with a single bird.

By about day eight to nine, when the wing feathers begin to unfold, the nest should be moved inside a hospital cage or suitable box cage for fledging and further rearing. Otherwise you will find, much sooner than expected, an empty nest on the table and no chick. It may have fallen from the table and gotten into trouble in the room, especially if you have four-legged pets sharing your home.

A hospital cage (chapter 3.2.2) with three sliding front panels is very useful for hand feeding a rescued chick: the panels can be positioned to give a small opening to feed the chick and to prevent it from flying or jumping out; extra heat can be given (chapter 11.2); high levels of hygiene can be provided; and perches can be arranged to assist the chick in developing its flying skills.

A single hand-reared chick needs contact for socialization, and thoroughly enjoys preening those areas of its body it can't reach for plumage and skin care. A watercolor brush is ideal for this. Sibling chicks begin allopreening once they fledge.

Weaning chicks from laborious hand feeding can be accelerated if the chicks are given the opportunity to explore and pick up suitable food soon after fledging. Drops of water moving over a leaf or other surfaces and moving insects trigger the first feeding response.

Small perch feeders can be made from pill bottle caps glued to clothes pins. One is used for water and one for food, such as cut up insect larvae and crumbs of egg cake. The feeders are clamped to the perch where the birds usually perch or roost. Curiosity and growing appetite, due to longer intervals of feeding by the foster parent, trigger a heightened interest in food. By using these perch feeders it is not unusual to see chicks become self-feeding as early as eighteen to twenty-one days of age. This is a week to ten days sooner than if the chicks are cared for by their natural parents.

Figure 7.5-3 Fledged chick investigating food in the perch feeder.

After about four to five days the chicks should be transferred to large cages to develop their flight muscles.

I create "feeding-sticks" out of approximately 16-inch (40 cm) long, thin bamboo twigs when I feed hand-reared, fledged chicks. Chicks quickly acquire a sense for a flight distance (chapter 3.3.1) and may not tolerate an approaching hand after they are released from a small hospital or rearing cage. Stick-feeding is also helpful for hospitalized birds and feeding special (medicated) food items through the mesh wire of an enclosure.

Stick-feeding waxworms and other tidbits ensures that each bird gets enough food after the fledged chicks have been released to a large enclosure when self feeding is difficult to monitor. Vitamin/mineral supplements or medication can be given this way by dipping waxworms into the supplement. If more than one chick is fed, it ensures that each get its fair share. The chicks continue to beg for food well past the point where they have become independent.

Figure 7.5-4 Stick-feeding fledged chicks after release into a large enclosure.

7.6 Weight Monitoring

Weight monitoring is the best way to track the development of a chick. It should be weighed at the same time of day, either before or after a meal, to obtain comparable results. Weight loss should not happen if all is well, but if it does, we need to be alarmed and find out the cause. At hatching time the chicks weigh .07 to .09 ounces (2 to 2.5 g). This weight may drop slightly on the first day, but then rises by about a gram or more per day until fledging when they could weigh 6.4 ounces (18 g). The following increased physical activity consumes energy and again there could be a loss of weight. A healthy adult weighs between .8 and 1.06 ounces (23 and 30 g).

An electronic letter scale is helpful for weighing birds in the aviary without capture. A mealworm dish is placed on the scale and the added weight of each bird alighting on it can be read off the digital display. Body weight is an excellent indicator of a bird's condition and progress.

Newly imported or resident birds can also be weighed this way and monitored over a given period of time.

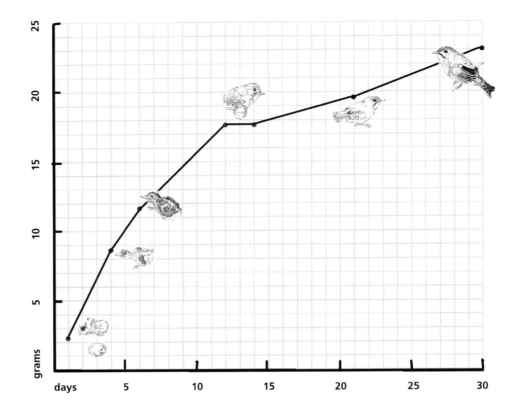

Figure 7.6-1 Average weight gain and development in hand-reared pekin robin and silver-eared mesia chicks of genus *Leiothrix*.

• Hatching weight is 2 to 2.5 g.
• Adult weight is 22 to 30 g.
• There is a typical slow-down of weight gain at fledging time and a flattening of the weigh gain curve as the chicks mature to adults.

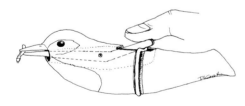

Fig 7.7-1 Hand-made pekin robin puppet with moveable lower mandible to hold food items.

7.7 Behavior Considerations

A further dimension of hand-rearing birds is the prevention of imprinting on the foster parent. This is more likely to occur if a single bird is raised.

Parent-raised birds develop natural behavior, which is one aspect of conservation of a given species. I created and used a pekin robin puppet to rear chicks as it is done with cranes and other species, which are very prone to imprinting.

The technique of puppet and mascot rearing is commonly done with captive cranes and raptors, where the eggs are removed for artificial incubation. In this method a look-alike model of the adult bird "cares" for the chicks until weaning age. I am not aware if this method has been applied to songbirds. Experimentally, I did raise two orphaned pekin robins this way, which were genetically important to the breeding program. The chicks were later introduced to other parent-raised birds without any behavior conflicts.

The preservation of species-specific song is noted in chapter (2.7). If the chicks are isolated from singing adults, they will pick up substitute songs and calls from other birds. These will be mixed into their innate song pattern. While this occurs in nature as well we need to be mindful that the chicks should be exposed to adults of their kin or recordings of their species-specific song. It would be counterproductive to raise pekin robins that sing like canaries. It is much simpler to obtain and care for a canary in the first place.

7.8 Hand Feeding Log

The following table is a page from a hand-feeding log of a pekin robin chick. The chick was eight days old and was taken in for hand-rearing due to banding the previous day, which had caused disruptions of parental care.

Time of Day	Items Fed	Defecation	Remarks
6:30 a.m.	3 mealworms 1 cricket belly	firm, encased in fecal sac	15 g body weight before feeding
7:00 a.m.	3 mealworms 1 waxworm	firm, in sac	vitamin/mineral added to waxworm
7:45 a.m.	3 moth bellies 1 waxworm	none	added calcium/vitamin D
8:30 a.m.	2 mealworms	small amount	refused a big cricket belly, try wild insects
9:30 a.m.	2 waxworms	none	not eager to feed, next feeding to be delayed
11:30 a.m.	3 spiders 2 cave crickets	greenish, no sac	appeared tired, eyes closed a lot
2:00 p.m.	2 cricket bellies 1 mealworm	good feces, normal color	
3:30 p.m..	2 mealworms 2 cricket bellies	none	preening wing feathers
4:45 p.m.	2 cricket bellies 2 spiders	good feces	
6:00 p.m.	2 spiders 2 crickets 2 mealworms	good feces	deposited feces on rim of nest
8:00 p.m.	3 mealworms 1 waxworm	good feces	
9:00 p.m.	1 waxworm 1 mealworm	no feces	was sleeping prior to feeding, 16 g body weight

Note: In this case a high number wild-caught insects were fed to stimulate the bird's feeding interest during the acclimation period.

If possible, "white" mealworms and crickets that have just shed their exoskeletons are selected for hand feeding chicks.

7.9 Addendum: Protocol for Rearing Chicks from Day 1

I have not explored systematic artificial incubation and hand rearing of day old chicks, but feel that one should have the capability to do so; particularly in cases where eggs or chicks of new founders are abandoned and chances to obtain their genetic expression becomes hopeless. I have lost one- to three-day-old chicks and suspected fluid aspiration or inadequate diet and husbandry to be the fault.

My colleagues at the San Diego Zoo led me to Alan Lieberman and Cyndi Kuehler, principal managers of the Hawaii Endangered Bird Conservation Program, a conservation outpost of the San Diego Zoo. It has been in operation since 1993 and has had remarkable success in artificially raising rare softbills from "day one". With permission, I am inserting this e-mail excerpt from Alan Lieberman's reply to my questions.

Q: What is the best food for newly hatched pekin robin-type softbills?

A: Bee larvae are magic. We get frames from the local apiarists and pick out the larvae (NOT the pupae!). We freeze them in Pedialyte® in an ice cube tray. Make sure all the larvae are covered with liquid. We usually top off the tray after one day of freezing to keep the delicate tissue from "burning". We never use pupae for nestlings because of the high level of chitin. We tried various trials of pupae with young house finches and killed a bunch with impaction.

Q: How soon should one feed the chicks and how much?

A: We feed chicks within hours of hatching. In fact, we have even fed chicks after they "capped" but were still sitting in the egg. Why not? What's the danger? I think there are a lot of myths and legends that must be hanging around aviculture as if Aristotle himself declared it so, so it must be.

I cannot comment on other people's formulae. I am not a nutritionist but I can only say what has worked for us. If there is something out there that works for others then I endorse that protocol for them. We have been criticized too often by people who insist their way is the only way. I only know the best way is the way that works. To each their own.

So we start all the chicks off with a daily ration of 25 percent by body weight of food. We feed every hour starting at 0600 (6 a.m.), ending at 2000 (8 p.m.). Count 'em. That's fifteen feedings. Take the 25 percent of the chick's weight at day one, divide that by fifteen, and that is EXACTLY how much you feed at every hour. So, for example, if a chick weighs 3 grams on day one, it will get .05 grams at each feeding: 3 grams X .25 = .75 divided by 15 = .05. That's the quantity.

We use three items on day one: bee larvae, cricket guts, minced hard-boiled egg. We add Pedialyte® for moisture. We alternate the three food items by the hour, starting with bee and ending with bee. We begin supplements of brewer's yeast, bone meal on day one, alternating every hour, and then add calcium into the rotation on day three (just a bare dusting like salting your eggs for breakfast). Day two goes to 30 percent, day three to 35 percent, etc. up to about 65 percent. It will get to a point where the chick can't eat its full ration, but by then, you'll know how much is good and how much is too much.

Q: When do you feed insects?

A: We start adding different items by day four, depending on the species. Items we feed are waxworms, mealworm guts only, papaya, Gerber's baby peas mixed with Gerber's baby oatmeal, pureed fruit cocktail, even frozen fruit flies for the insectivorous spp. But these items we work in slowly over time.

Q: How is temperature regulated?

A: Temperature. We start at 98°F (37°C) for day one, and decrease by a degree a day until they hit the mid 70s (24°C). By then they should be feathered and at room temperature. Humidity is a problem, but we keep a dish of water in the brooder at all times, and try to always keep the relative humidity in the 70s.

Q: How do you keep them hydrated?

A: We NEVER give oral fluids. Aspiration for sure. We add fluids by making the food plenty moist with Pedialyte®, or Nekton Lory or Nekton I. Be careful with the latter two items. They're potent and can over-tax a very young bird. We generally go with Pedialyte® for the chicks that are less than four days old.

(The protocol has been principally developed by Cyndi Kuehler of the San Diego Zoo.)

— ALAN LIEBERMAN
18.8.2004

It is so encouraging to have a helping hand from colleagues, scientists, and aviculturists who work in this field.

— PK Jan 5, 2006

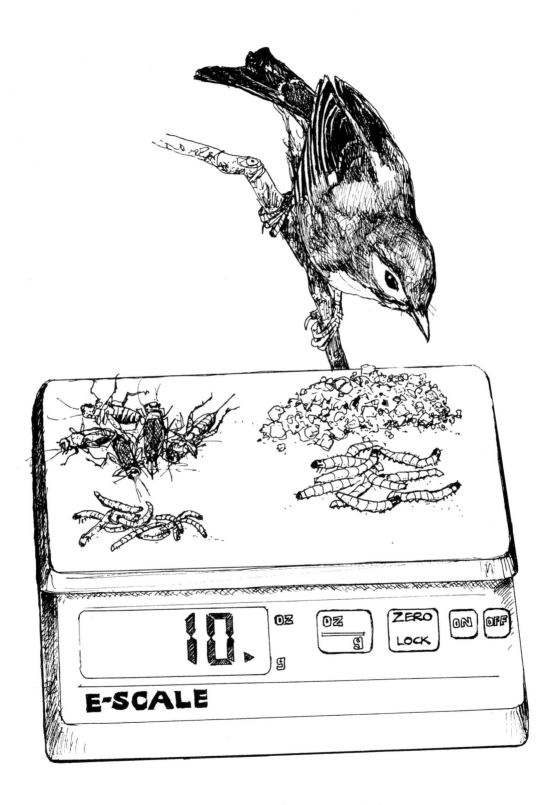

Figure 8.0-1 Daily food aggregate for an adult pekin robin.

Feeding

Note: measurements in nutrition and medicine use the metric system.

Feeding adult pekin robins outside their breeding season is simple. The insectivore–frugivore orientation in their natural diet, which includes, besides insects, some soft plant tissues and certain accessible seeds, allows us to provide them with commercially prepared food. This is usually sold as softbill or universal softbill food preparations in the form of dry crumbles or pellets.

Complete diets in pelleted or cubed form have become the norm in animal husbandry situations for agriculture, pet care, as well as zoological gardens. It takes the worry out of unbalanced diets. In flock feeding it eliminates concerns about the most dominant animal(s) hogging the choicest items and leaving the subordinates with inferior leftovers. Theoretically we only need to offer clean water besides such balanced and complete food. Commercially raised insects are readily available to become welcomed occupational food. Good quality softbill food can be purchased from pet food suppliers, which could end the discussion on maintenance diets at this juncture. But even for maintenance diets, there are trends to swing back to more "close to nature" food offerings, primarily to stimulate feeding-related behavior responses.

Feeding pekin robins in a breeding program is hugely different. The fundamental difference rests with the provision of high quality insect food and sufficient quantity of live insects to rear the young. Correspondingly, the chapter on propagation of suitable species of insects and their preparation to ensure an effective food value has been given special attention (chapter 9).

We can easily find good references in animal husbandry literature on nutrition. However, to develop an understanding of the subject, or simply to refresh the reader's knowledge, some basic principles are reviewed here.

8.1 Water

A simple concept of nutrition is that it takes a defined blend of organic and inorganic elements to build a certain organism. An organism can be analyzed by measuring its composition of water and dry matter. A bird is composed of roughly 70 percent water and 30 percent dry matter. If we were to burn it, the organic substances would be consumed and we would be left with ashes, composed of inorganic minerals and trace elements.

Water is obtained by drinking and also through the breakdown (metabolism) of organic compounds, which contain water. Fresh fruit can provide a good source of water, which is why we add it to a shipping container as a back up in case the water is spilled or not available. The loss of only 10 percent of body water is a serious problem for a bird. A bird can easily perish within twenty-four hours without access to water in one form or another. A bird needs the equivalent of 50 to 60 ml/kg body weight (BW) of water for daily maintenance. For an adult pekin robin that is approxiamtely 1.5 ml/day.

8.2 Protein

Proteins are composed of twenty-two different amino acids. About half can be produced within an animal; the remaining amino acids must be obtained from external sources. A bird needs ten essential amino acids to build its body protein. The combination (types and proportions) of the amino acids determines the quality protein in a diet. The closer the match the better.

Individual plants have different, mostly incomplete, combinations of amino acids. The soybean is the exception and although it is low in the amino acid methionine, it has a good balance of the other essential amino acids. In comparison to most plant proteins, animal proteins have a good balance of the essential amino acids. A bird egg is a complete package. It has complete protein and all the other nutrients that are required to "build" birds. This is why I find it a great relief when the chicks begin to consume egg cake.

Protein yields as much energy as carbohydrates, but its conversion into energy is not as efficient as compared to carbohydrates and fat. I target crude protein levels of 16 to 18% in the softbill diet (chapter 8.11) for maintenance. The provision of insects raises the protein ratio (chapter 8.13) for the breeding season. If the animal must break down protein contained in muscle and other body tissue to stay warm, it begins to lose weight. Provision of more protein in a diet to provide fuel is not practical, since extended over-supply of protein may put stress on the kidneys and liver (chapter 12.2.5). Good sources of animal protein are eggs, cottage cheese, beer yeast, puppy and cat pellets, and insects (chapter 8.13).

8.3 Carbohydrates

Carbohydrates supply energy. Carbohydrates are compounds including sugars, starches and cellulose. All are built from carbon, hydrogen and oxygen. Glucose has a simple sugar unit (monosaccharide), also known as dextrose, fructose or grape sugar. Beet sugar (household sugar) and starch are built from monosaccharides in succession and water molecules must be removed for each step.

Cereal products, seeds, vegetables and fruit, as well as honey, are sources of carbohydrates for avian species.

Because of an interrelationship between fat and carbohydrates, deficient carbohydrate levels in the diet deplete fat reserves to create energy and conversely oversupply can result in obesity.

A complex carbohydrate is cellulose; birds cannot digest it unless they are equipped with organs (cecae) that permit bacterial action to do so. Grouse and ptarmigan are examples of species with this gut modification, which permits them to feed on buds during the winter. Geese also have large cecae. Many plant-eating mammals, like horses and cattle, get this help from mutualistic bacteria in their digestive tract, which produce enzymes to ferment cellulose into digestible forms of carbohydrates, or volatile fatty acids that are absorbed for energy.

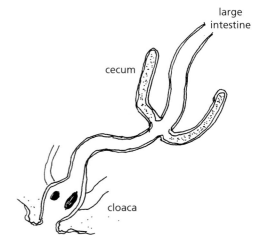

Figure 8.3-1 The cecae of a grouse.

To a songbird, cellulose is roughage or fiber. It creates bulk and helps the gut action (peristalsis) to move the food mass through the system. The more fiber a diet contains, the firmer the feces. The dry and firm feces of a seedeater forms due to the insoluble fiber in the seed coat, but also from the soluble fiber (starch) in the seed itself. The hard outer body (exoskeleton) of an insect also works as indigestible roughage; it is comprised of a substance called chitin, which physically and chemically resembles cellulose.

The feces are a good indicator of a bird's digestion. Unexplained changes must be investigated. For example, the feces of a seedeater will change its consistency if large amounts of fruit or greens have been consumed. Fruit-eating birds' feces normally show a loose consistency.

Birds do not produce urine like mammals. The white urates secreted by the kidneys, which is comparable to urine in mammals, should be "side by side" with the feces, which are the dark, undigested food particles. Pekin robins usually have soft pasty droppings.

8.4 Lipids

Lipids are fats and oils, which are broken down into fatty acids in the bird's gastrointestinal tract. There are essential fatty acids a bird needs, but they are rarely absent from a varied diet.

Fatty acids are needed for the absorption of fat-soluble vitamins and certain body functions, especially energy storage. They provide more than double the energy (calories) per weight unit than protein and carbohydrates. Fat reserves are winter fuel for birds when the food supply is irregular. Migratory birds could not migrate and penguins could not breed, without their fat reserves.

Carbohydrates can be converted into fats, which in turn become fat deposits to store energy. Metabolism reverses the process and provides sugars for the muscle action and body heat plus water without solid waste products. Fat is the perfect fuel for birds during migration when they do not make frequent landfall.

8.5 Vitamins

Vitamins are either water-soluble or fat-soluble. Vitamins C and the B-complex group are water-soluble and, as such, not stored in the body, which means they must be offered to the birds on a regular basis.

Pro-vitamins are substances that promote the formation of vitamins by an animal. Besides naturally occurring vitamins in the food constituents, we can also provide synthetic vitamins, which are sold in high concentration by various manufacturers. They must be used with diligence to stay within proper dosages.

B-complex vitamins are the catalysts to life. A multitude of cell functions depend on enzymes, which are activated by these vitamins. B-vitamins are found in high concentration in the germ of a seed, sprouting seeds, and growing leaves. In animals, B vitamins are most concentrated in the tissues of kidney, liver, brain, and muscle. Yeast is another source.

Healthy birds can produce vitamin C in their bodies, but not store it adequately. Vitamin C is, therefore, supplemented at times of high stress to strengthen the immune system.

Vitamin A, D, E, and K can be stored with the bird's fat. Without replenishment they deplete slowly to create a problematic deficiency, called hypovitaminosis. The storage capability has a drawback, however; an over supply (hypervitaminosis) of vitamin A can occur and cause toxicity, while calcinosis of the aorta and kidney can be aggravated through over supply of vitamin D. Hypervitaminosis for E and K are less of a concern.

Vitamin A is important for healthy mucous membranes throughout the body to ward off a host of infections and gout. It supports development of the embryo, bone growth, and immune systems. Pro-vitamins (in this case, carotenes) are converted to vitamin A in the bird's liver. Sprouted grain, dandelion, chickweed, carrots, nettles, and spinach are good sources of carotenes. Egg yolk and liver oil contain vitamin A, which can be utilized directly by a bird.

The vitamin D group, D_2, and especially D_3, are vital catalysts to mineral absorption for bone formation. Deficiency leads to low fertility and poor hatching success because the embryo cannot adsorb calcium from the eggshell for its development. The vitamin D group also includes pro-vitamins, which are converted to highly effective forms of vitamin D by the UV rays of the sun on the surface of the skin and feathers involving body oils. Birds that are not exposed to direct sunrays may need vitamin D_3 supplements. New acquisitions that have been in long-term indoor quarantine facilities may lack vitamin D_3. Vitamin D_2 is present in green plant tissue. Vitamin D_3 is also found in fish liver oil and egg yolk.

Vitamin E supports metabolism, nerve functions, and sperm production. It is found in sprouted seeds, flax, and corn.

Vitamin K supports coagulation of blood and enzyme functions. It, too, is found in sprouted seeds and green plant tissue.

The amount of vitamin and mineral content in a diet is not easily established for the average bird keeper. Only detailed analysis of all food constituents could reveal the levels. Typically vitamins are listed in International Units (IU); milligrams (mg) which is one-thousandth of a gram; or micrograms (mcg or µg) which is one millionth of a gram. Minerals are presented in milligrams (mg) per kilogram, or as percentages.

Some references note daily requirements for a given animal species for vitamins and minerals per kilogram of body weight, while others suggest provisions per kilogram of food mass offered to the animal, which is how most requirements for domestic poultry (our best model species) are displayed. Setting requirements per species will vary with intake levels, individual body size, and metabolic rates, whereas in diets that have been formulated as a concentration, amounts of specific components will vary accordingly with intake levels. Based on the requirements established for a variety of poultry, then, a nutritionally balanced maintenance diet might be expected to contain the following per kilogram food mass:

Vitamin A	8,000 – 10,000 IU
Vitamin D	800 – 1000 IU
Vitamin B_{12}	10 – 20 mcg (µg)
Vitamin B_1	1 – 2 mg
Vitamin E	80 – 120 mg
Vitamin K_1	0.25 – 0.5 mg
Vitamin C	1g — only in situations of high demand
Calcium (Ca)	5 – 10 g
Phosphorus (P)	2 – 4 g

To extrapolate daily provisions for an individual bird based on kg food mass, one must establish how much food a bird consumes.

Rough calculations suggest that a 25 g bird (a pekin robin) consumes half its bodyweight in food per day, on average 12 g of food. 1000 g food mass divided by 12 g = 83 portions. Daily provisions in this theoretical model would then be 1/83, i.e. Vitamin A 96–120 IU/day per bird.

If concentrated supplements are added to the diet they must be balanced with the levels of vitamins and minerals in the basic diet. Sufficient nutrients are necessary for breeding, growing, dealing with stress, and molting. Higher fat or carbohydrate levels are helpful in the cold season, but increased daily rations in volume are usually sufficient to provide more energy to keep up body temperature and condition.

8.6 Minerals and Trace Elements

Minerals are essential for body construction and functions. Calcium deficiency associated with a lack of Vitamin D_3 is of particular concern in the rearing of chicks (chapters 6.0 and 11.1.7).

Growing animals, or molting and reproducing animals, have higher requirements than adults in a maintenance mode. Building muscle, bone, feathers, and other tissues requires more nutrients than replacing aging cells or maintaining the basic body functions. This is one reason why breeding pairs and their growing chicks need different diets.

When a hen pekin robin lays four eggs at 3 g each, she produces building blocks equal to half her weight in four to five days. One can easily comprehend why it is very draining on the hen if she loses her clutch of eggs and re-lays, or if she is made to raise too many broods consecutively.

We should not be misled by highly bred strains of layer-chickens. It has taken hundreds of years to select those hens that lay more eggs and survive the drastic demand on their system. Besides, layer-chickens are "spent" after about two years and have a poor record of longevity. A hen is born with a finite number of undeveloped ova in her ovaries. Once they have been laid the bird becomes infertile, which is one of the reasons why high-performance layer-chickens are replaced at a very young age.

For maintenance, a concentrated vitamin/mineral/amino acid supplement is added to the daily food mix, in amounts below the producer's recommended maximum rate. This is more as an insurance policy rather than based on detailed supply and demand calculations, because these constituents are already in the basic food mix. Birds under stress and with suspected deficiencies, of course, receive elevated rations. There is a margin of error the bird tolerates at least for a limited time period.

Twenty-two minerals are divided into macro and trace elements. Calcium, phosphorus, magnesium, potassium, chlorine, and sodium, for example, fall into the macro group. More than 50 mg per kg body weight are found in an animal, while trace elements are of lower concentration.

Iron storage disease has been experienced with frugivores, which build up unfavorable concentrations in the liver. Less than 40 ppm (parts per million) Fe is the goal in a balanced diet for softbills; 60 ppm is noted as the upper limit by some aviculturists.[1] Dried fruit can contribute to iron storage disease, as well as dry commercial products, because they contain higher concentrations. Excess vitamin C can also be problematic in increasing iron uptake.

I have not encountered this problem in the genus *Leiothrix* that are cared for on the diet discussed in this text. The as-fed diet of dry mix and egg cake (8.11 and 8.12) has been reviewed and it contains 38 ppm Fe.

Minerals are found in the various plant and animal foods. In aviculture sterilized, finely crushed eggshell or oyster shell and proprietary mineral supplements are used to boost diets. I offer crushed eggshell free choice in special dishes where rain and feces do not mix with the contents. Calcium and phosphorus should be represented in a diet with a relationship of 2:1 to 1.5:1. Other macro elements are generally present in sufficient amounts in mixed diets.

Trace elements are required in minute amounts and normally supplied via the ingredients of a basic diet. They are also present in the soil of the planted outdoor aviary. Comprehensive, proprietary, vitamin/mineral supplements contain needed trace elements as well.

8.7 Food Presentation

Feeding your birds entails more than providing adequate and balanced nutrients. The food must be accessible, consumed and digested to give the bird what it needs. Effective food presentation is very important to achieve this.

Pekin robins can survive on dry food pellets and water, but this is not ideal if the object is to bring this species into top condition to breed and to live a long life. Since digestive enzymes and other body chemicals are dependent on fluids to function, it is beneficial if food arrives in the stomach already moistened, to make digestion easier and more efficient.

Seedeaters have a crop to pre-moisten dry foods. Pekin robins do not. The dry food mix can be moistened with grated carrots and

---

1. Theo Pagel, pers. com. 2006.

apples or their juice, as well as cottage cheese, prior to feeding to make it more palatable. The dry food mix can be prepared and refrigerated for a few days in advance, since it takes a few hours for the liquids to saturate the mix.

The food mix should only be slightly moist and still crumbly, but not soggy. Fresh or freshly thawed fruit, such as strawberries or blue berries can be added to the prepared mix on a daily basis. Egg cake (8.12) is diced and added to the mix. In my experience, all of the egg cake is consumed with regularity and therefore can be used as a vehicle to supply a vitamin-mineral mix to the birds by dusting it onto the cake slices before stacking and cubing them. The powder adheres well to the moist cake and little or none is wasted.

On average each pekin robin receives about 10 to 12 g of total food for maintenance. I like to see a small amount of food left to ensure adequate food supply. The leftover food is recycled to the insect colonies.

Insects are fed to the birds daily in small amounts, about 2 g per bird. Mealworms and waxworms are offered year round. Most pekin robins show little or no interest in crickets outside the breeding season.

The food volume is increased in cold weather by up to 20%, depending on consumption.

8.8 Feeding Utensils

Although pekin robins frequent the shrub level in their habitat and feed, at times, on the ground in search of insects, they prefer their food station to be raised off the floor. It gives them better access for their method of taking food by the grab-and-run method. It is safer and easier to fly off from an elevated position than from the ground. Newly introduced birds will accept the presented food with less hesitation, if they can access it at shrub level. This also makes it much more convenient to the keeper, avoiding bending to the ground to exchange food dishes. The contamination through debris from the aviary floor is prevented as well. One of my friends also insists that there is great value in having the food dishes in plain view to monitor consumption and that feeding has not been forgotten. A very practical and safe feeding station, which can be mounted on the aviary front, is described in the chapter on Housing (3.3.8.). It should have room for at least two dishes and a slice of orange or other fruit. One dish is used for the food mix and the other to hold mealworms. Glazed ceramic dishes, which are sold for pet hamsters, of about 3.5 ounces (100 ml) capacity hold 1.06 ounces (30 g) of the above-described food. This is enough for three birds. If more birds are fed in one aviary, additional dishes are used.

Figure 8.8-1 Food station is positioned above ground for "grab-and-run" feeding.

The fruit slice is speared on a short stainless steel screw, which protrudes through a heavy plastic block. This anchors the fruit to help the birds to pick the flesh away from the peel. It prevents the tray from becoming messy and avoids the need for the daily washing of the tray to curb fermentation or mold. The plastic blocks and the dishes are exchanged with a back up set each day for cleaning. This adds great efficiency in serving the aviaries and affords the necessary degree of hygiene.

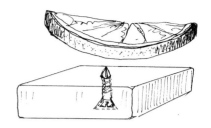

Figure 8.8-2 Plastic block to hold fruit.

Mealworms are unable to escape from the glazed dishes, but waxworms and crickets can. These insects must be killed before feeding or fed directly to the birds. Crickets can be offered in plastic storage boxes, which work as self-feeders (chapter 6.10).

A small dish with crushed eggshell and a mineral supplement is always available in each aviary. It is positioned near the feeding station. While these birds do not consume much of it, a few particles are found when necropsies are performed on perished birds, including very young chicks.

The feeding station must be shielded from bird excrements. A piece of stiff clear plastic is suitable. It allows the birds to see both the food from above and another bird at the station, plus it is easy to clean. Most pekin robins are rather polite and wait until the other bird has left before alighting. The same type of cover can be installed over a birdbath.

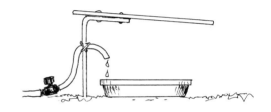

Figure 8.8-3 A debris shield over a birdbath.

8.9 Drinkers

The provision of clean water must be done without fail. While the birds may have access to water in their birdbath, it is prudent to always offer water at the food station. This is done with standard watering tubes (chapter 3.3.8). The water is made more palatable by adding 10 to 20 percent sugar per volume, to entice the birds to drink it rather than their bath water. The tubes are washed and replaced daily, and more frequently in hot weather. One can add carrot or fruit juice to it, but fermentation is then accelerated.

I have observed parents deliberately dip a piece of egg-cake into the drinker to transfer water to their nestling in hot weather and, to my even greater amazement, I saw them dip the moistened food into the mineral supplement to transfer it to the chicks.

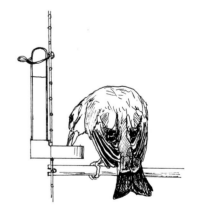

Figure 8.9-1 Drinkers encourage birds to ingest clean water.

8.10 Feeding Routine

Softbill birds must be cared for with more rigid routines than seedeaters. Food hygiene is of particular concern, inasmuch as the moist substances are prone to spoil in warm weather and to freeze in cold weather.

This group of birds is very active for most of the day, has high-energy requirements, and feeds on the go without a crop to store food. The high activity is necessary to hunt insect food that is mobile and much more dispersed than ripe seeds in concentrated patches of plants.

Softbills have a high rate of metabolism. A long night makes them very hungry at daybreak, which is the reason why we need to feed them as early in the morning as possible. The short daylight season is problematic in that respect. Bird fanciers who like to sleep in until the sun is high in the sky are better suited to keep finches and hookbills. Birds adjust to scheduled routines and benefit physiologically from being fed at the same time, just as we do.

In hot weather we may find that fermentation of moist food and sugar-water may set in before the end of the day. It may become necessary to feed again in the late afternoon and/or to remove the food dishes in the evening to prevent ingestion of spoiled food first thing in the morning when the birds are most hungry.

8.11 Diet Information

There are as many diets as there are pekin robin keepers, which points out how adaptable the species is to a wide variety of food constituents. For maintenance, the "as-fed" staple food mix for insectivores should have approximately 18 percent protein. During reproduction and active growth the protein level can be increased to 20 percent. Rising protein supply triggers reproductive activity, hence the relationship between emerging insect population and nesting by insectivores. This has application in animal husbandry to boost fertility by mimicking nature and feeding more insects. Crickets have about 20 percent protein, while mealworms have close to 19 percent, and waxworms about 14 percent (8.13).

Protein is measured as crude protein in a food analysis. Not all of it is digested and absorbed. The protein level of food insects is influenced by the food given to them. Their gut nutrient content can vary based on how they have been cared for. Food insects that have been starved have lower values.

We will investigate how to improve the food value of insects in chapter 9.

Following is a diet that has produced consistently good results. The diet has three components:

a) dry mix, which can be prepared in large quantities and stored in a freezer (recipe and analysis below),
b) processed carrots, fruit, and cottage cheese to moisten the dry mix,
c) egg cake (recipe and analysis follows [8.12]).

Component (a) is used to create a moist crumbly diet. It could be freshly prepared for perhaps three to four days and refrigerated until offered to the birds. Large batches of premixed food are stored in a deep freezer to extend shelf life.

The diet components (a) and (b), are mixed in equal volume amounts and fed daily. The 50:50 combination of the two components represents a satisfactorily balanced diet. The dry mix or the egg cake would need to be adjusted if fed on their own. During the winter months more of the high-energy egg cake is offered for consumption by the birds.

A laboratory has specifically analyzed samples of the dry mix and the egg cake. Results are included with the recipes, below.

Dry mix recipe by volume:

- 20 parts softbill composite pellets (18% protein)
- 20 parts puppy chow (20 to 25% protein, <12,000 IU vitamin A per kilo)
- 4 parts oatmeal
- 4 parts muesli or Grape-Nuts breakfast cereal
- 4 parts poultry layer pellets (16% protein)
- 2 parts chopped sunflower seeds
- 2 parts powdered whey
- 1 part powdered yeast
- 1 part chopped hazel nuts/almonds
- 1 part chopped dried fruit (papaya)
- 1 part poppy seed
- 1 part wheat bran
- 1 part wheat germ
- 1 part soybean protein powder
- 1/2 part bee pollen

Coarse ingredients should be fine-ground in a coffee grinder.

Dry mix analysis (as fed):		Vitamin values*	
Ash	5.74%	Vitamin A	9,440 IU/kg
Moisture	7.60%	Vitamin D$_3$	580 IU/kg
Crude protein	21.40%	Vitamin E	141.63 mg/kg
Fat	10.34%	Vitamin B$_{12}$	10 mcg (µg)/kg
Acid detergent fiber	5.00%	Vitamin B$_6$	1.91 mg/kg
Calcium	1.30%	Vitamin C	14.05 mg/kg
Phosphorous	0.85%	Biotin	0.01 mg/kg
Potassium	0.59%		
Magnesium	0.19%		
Sodium	0.14%		
Salt	0.36%		

*Approximate vitamin values have been established by Ellen S. Dierenfeld Ph.D., Saint Louis Zoo in Missouri, USA, via a food constituent database "Zootrition 2.5."

Egg Cake Recipe

Preheat oven to 350°F (175°C).

Line four loaf pans (or two larger ones) approximately 3¼ x 7 x 2¼ inches high (8 x 18 x 6 cm) with wax paper.

Ingredients:

1½ cups (375 ml) 2% salted margarine

1½ cups (375 ml) sugar

6 eggs

1½ cups (375 ml) whole-wheat flour

¾ cup (175ml) soy protein concentrate (from health food stores)

1½ teaspoons (7.5 ml) baking powder

¼ cup (70 ml) gluten

Instruction:

- Cream the margarine and add sugar.
- Beat until very fluffy and light (this is important).
- Add eggs, one at a time and beat well after each one.
- Add dry ingredients and divide batter between loaf pans.
- Bake for 50 minutes or until done.
- Take out of the oven and leave the pans for about 10 minutes.
- Take loaves out of the pans and cool on racks with wax paper left on.
- Let cool and feed, or freeze for future use.
- The loaf in use is kept in a refrigerator in a plastic bag to retain its moisture.

8.12 Egg Cake

The egg cake is sliced and finely diced and enriched by dusting it with vitamin/mineral supplement. The cubes should be cut small enough to allow the birds to swallow them whole; otherwise the birds will break them up by holding them on the perch with their feet and picking them apart, causing pieces to be dropped, which wastes food. Egg cake is mixed with the dry mix 1:1 to per volume parts, with higher proportion of egg cake, 1.2:1, during near freezing weather. Breeders can produce egg cake economically in their home kitchens. A cup of poppy seed can be added to entice seedeaters to accept egg cake.

Egg cake analysis (as fed):		Vitamin values*	
Ash	1.55%	Vitamin A	15,000 IU/g
Moisture	18.80%	Vitamin D_3	450 IU/kg
Crude protein	12.80%	Vitamin E	51.64 mg/kg
Fat	27.40%	Vitamin B_{12}	0.00 mcg/kg
Acid detergent fiber	1.60%	Vitamin B_6	1.53 mg/kg
Calcium	0.11%	Vitamin C	0.57 mg/kg
Phosphorous	0.23%		
Potassium	0.13%		
Magnesium	0.04%		
Sodium	0.38%		
Salt	0.96%		

*Vitamin values as calculated by Ellen S. Dierenfeld Ph.D.

In general insectivores consume more fat than seedeaters. Mealworms have approximately 14 percent fat and waxworms 25 percent, most grains have less than 5 percent; however, sunflower seeds (shelled), nuts and flax seeds have well over 30 percent on an as fed basis. Fat makes food very palatable and provides high energy. The main concern in providing a fat rich diet is the risk of obesity, with its related problems.

The above egg cake has been fed for several years and gross examination on all deceased birds is performed. The fat deposits were normal for birds, which died through accidents and, thus, in healthy condition. I conclude that the relative high level of plant-based fat in the above diet has evidently no negative effect with this highly active species if it is maintained in spacious outdoor environments. Longevity and fertility of the flock has been good on the above diet.

8.13 Food Value of Insects

The review of the food value of various insects explains why feeding some species in excess, and exclusively, can lead to deficiency and imbalances, for example, the lack of calcium in mealworms can cause rickets and conversely, the high fat content in waxworms can lead to obesity. Over supply of protein can lead to gout, particularly in older birds (chapter 11.2.4).

Food analysis of live insects (as fed)*					
	Adult Crickets	Cricket Nymphs	Mealworms	Waxworms	Buffalos
Ash (%)	1.1	1.1	0.9	1.2	1.46
Moisture (%)	69.2	77.1	61.9	58.5	63.0
Crude protein (%)	20.5	15.4	18.7	14.1	21.6
Fat (%)	6.8	3.3	13.4	24.9	10.6
Acid detergent fiber (%)	3.2	2.2	2.5	3.4	5.0
Calcium (%)	0.041	0.027	0.017	0.024	0.02
Phosphorous (%)	0.295	0.252	0.285	0.195	0.23
Ca:P	1:7.2	1:9.3	1:16.8	1:8.1	1:11.5
Vitamin A (IU/kg)	217	156	301	57	not tested
Vitamin E (IU/kg)	22	24	11	194	not tested

*The data for adult crickets, cricket nymphs, mealworms and waxworms were provided courtesy of Mark D. Finke Ph.D., Scottsdale, Arizona, USA. Data on buffalo worms were established through a commercial lab test initiated by the author February 2005.

The inverted Ca:P relationship points out the importance of supplementing Ca when feeding the above live insects to growing birds. The lack of calcium causes rickets, with clinical signs of lameness, deformed beak, weak and deformed or broken bones.

"Gut-loading" insects is an important technique to improve the food value of insects. Dr. Mark D. Finke, nutritionist and entomologist, Scottsdale, Arizona kindly provide this information on "gut-loading:"

> Insects are an important food source for many insectivorous birds but, unless they are treated, insects contain inadequate levels of several nutrients (most notably calcium and to a lesser degree vitamin A) to meet the animal's requirement. The most common method used to help correct these deficiencies is to feed insects a special diet to alter their nutrient content (especially with regard to calcium) to make them a more complete diet for insectivorous birds and other species. Often called "gut loading" diets, they increase the calcium content in the insects, not because the insects absorb the calcium but rather because of the food retained in their gastrointestinal tracts. These high calcium diets

are usually fed 24-72 hours prior to using the insects as food, since high calcium diets are not suitable for the long-term health of the insects. A palatable, finely ground dry diet containing 7-9 percent calcium should be suitable for gut loading most species of insects. This can be accomplished by adding 20 percent calcium carbonate (40 percent calcium) to the insects' base diet.

Another method sometimes used is to "dust" the insects with a powder (typically composed primarily of calcium carbonate) just prior to being used as food. While effective, this method can provide variable results since the amount that adheres to the insects depends on the characteristics of the powder (calcium content, particle size and electrostatic properties), the size of the insects, and the ability of some insects to groom themselves and remove the added calcium.

8.14 Color Food

The color of feathers is due to pigments embedded into them and special cells structures or oil layers that diffract light to create iridescent colors. Pigments create the colors in pekin robins, while the blue in the blue-winged minla is created by structure. This becomes evident when a blue-winged minla is photographed with an electronic flash, which makes the blue much more splendid than seen otherwise.

The coloration of many wild birds fades in time when they are maintained in aviaries. Wild pekin robins show a rich olive green, yellow and red hues. Fat-soluble carotenoids and lipochromes are responsible for red, orange and yellow pigments, while melanins create black, grey and brown colors.

Proprietary color foods are on the market to enhance the intensity of plumage color. Synthetic coloring supplements such as canthaxanthin — which keeps flamingos and ibises glowing red in zoological gardens — is also available for cage birds. Some of these products are problematic for the bird's health if given in high doses.

Regular addition of grated carrots or carrot juice in the diet is believed to improve plumage color; however, I have not detected measurable effects by offering it. I have tried beta-carotene capsules, a product produced for humans as a vitamin A supplement, and mixed the oily content with margarine to dilute the concentration. It can be spread very lightly on egg cake slices.

These additives are only effective if provided from the onset of the molt to its completion, when the pigments can be built into the growing feathers.

I found that pekin robins maintain good color if they are kept in planted outdoor aviaries where they can find suitable green plant tissue (chlorophyll), have plenty of room to exercise, and are exposed to seasonal climate changes.

8.15 Fruits and Vegetables

Pekin robins, especially silver-eared mesias and other softbills, interact with the live, green foliage in their aviary by nipping off leaves and tender shoots. I have noticed this activity in particular during winter months. Tits dismantle bamboo shoots and plant buds in search of insects and their eggs. Some plant tissue is likely consumed this way. I have offered chopped lettuce, dandelion *Taraxacum officinale*, and common chickweed *Stellaria media* in food dishes but the greens have been of no real interest. The flesh of fruit and berries is eaten. Individual pekin robins differ in their acceptance of and preference for various fruit. Most birds in my flock generally favor oranges, blueberries *Vaccinium* spp., and red huckleberries *Vaccinium parvifolium*. I observe pekin robins picking salal berries *Gualtheria shallon* from bushes in their enclosure and carrying them around in their beaks. Some berries are eaten; others are used more as play objects.

Figure 8.15-1
Red Huckleberry *Vaccinium parvifolium*.

There appears to be a seasonal change of interest in fruit. During the onset of the breeding season and during chick rearing, very little fruit is eaten, perhaps replaced by greater interest in insects in order to obtain a higher level of protein in the food mix. However, hot weather, even though during the breeding season, leads to higher consumption of oranges. Slices of oranges are presented on plastic blocks (made for the purpose) and held in place with pins (figure 8.8). Frozen and thawed blueberries can be offered year round. More fruit is eaten in the fall, which mirrors feeding behavior in the wild when the birds are trying to build up fat for the winter, and insect populations diminish.

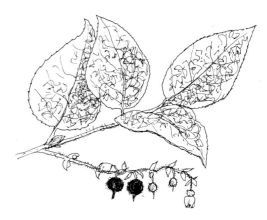

Figure 8.15-2 Salal *Gualtheria shallon*.

Huckleberries or pieces of red strawberries catch the attention of fledged chicks, if offered in a perch feeder (figure 7.5-3). They aid self-feeding through play and curiosity.

Apples, pears, peaches, papaya and strawberries can be offered. I find the latter is more readily consumed, in a grated form and used as a moistening agent for the dry food mix (chapter 8.11). A birdkeeper may try many fruits and vegetables not listed here. Pekin robins will pick through food offered to seed eaters and clearly eat some seeds on occasion. I mix chopped nuts, poppy seed and sunflower kernels mixed with suet for my various tits *Parus* spp. and European robins and see that it is regularly investigated by pekin robins.

Figure 8.15-3 Half apples or oranges can be offered on a branch in the aviary.

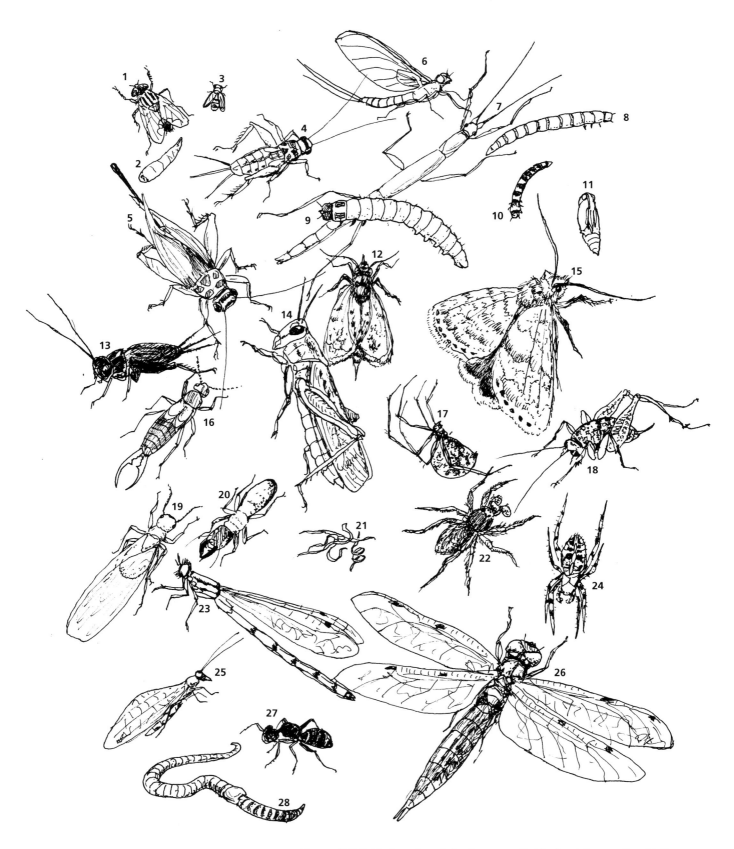

Figure 9.0-1 Various invertebrates are suitable food for small softbills.

9

Cultivation of Live Food

Key to Figure 9.0-1

1. house fly
   Musca domestica
2. maggot of house fly
   M. domestica
3. fruit fly
   Drosophila sp.
4. house cricket, male
   Acheta domesticus
5. house cricket, female
   A. domesticus
6. mayfly
   Family Heptageniidae
7. stick insect
   Carausius morosus
8. mealworm
   Tenebrio molitor
9. waxworm
   Galleria mellonella
10. lesser mealworm
    Alphitiobius diaparinus
11. mealworm, pupa
    T. molitor
12. waxmoth
    G. mellonella
13. field cricket
    Gryllus sp.
14. grasshopper
    Family Acrididae
15. owlet moth
    Family Noctuidae
16. earwig
    Forficula auricularia
17. house spider
    Achaearanea sp.
18. cave cricket
    Ceuthophilus sp.
19. dampwood termite, winged adult
    Zootermophis angusticollis
20. dampwood termite, soldier
    Z. angusticollis
21. whiteworm
    Enchytraeus albidus
22. wolf spider
    Family Lycosidae
23. bluet
    Enallagma sp.
24. orb-weaver
    Araneus sp.
25. lacewing
    Chrysopa sp.
26. skimmer
    Libellula sp.
27. carpenter ant
    Camponotus sp.
28. earthworm
    Lumbricus sp.

The perception that it is difficult to breed insects, and the notion that the breeder must visit his aviaries several times per day to provide live insects to the chick-feeding parents, has deterred people from becoming softbill breeders. This chapter is meant to allay those concerns by presenting an approach to simplified insect breeding and feeding. A key issue is feeding insects properly so that they, in turn, provide the birds with their required nutrition. A food insect is as good as what it has eaten. I recycle left over softbill food to the various insect cultures as one step towards that end.

In softbill aviculture "live food" embraces any live, whole animal used as food, from tiny fruit flies for hummingbirds, to small mammals such as mice or poultry chicks for hornbills. The latter are usually killed before feeding.

For the species addressed in this book, primarily certain insects and spiders (Arthropods) and, to an extent, segmented worms (Annelida) are of interest. Spiders are not cultivated, but they can be collected to provide some tidbits, which are highly cherished by pekin robins feeding their chicks (chapter 6.11).

We need to be cognizant of misnomers such as mealworms and waxworms, which are not actual worms but larvae of insects and, as such, need a lot more air circulation and significantly less moisture in their cultures than segmented worms.

Insects undergo metamorphosis in their development from egg to the reproducing adult form (imago). Crickets, stick insects, locusts, and grasshoppers have a simple metamorphosis. The hatchlings (nymphs) look like wingless adults and change little as they grow. Mealworms, waxworms, silkworms, and butterworms are larvae of beetles and moths, which change into pupae, which then hatch into adult insects. This is a complete metamorphosis that also applies to flies.

Some commercially traded larvae are treated with radiation to arrest metamorphosis in the larval stage for extended "shelf life" and import regulations. Insects treated in this way cannot be used to start a culture for propagation. Hormones are used to create "giant mealworms;" likewise, these will not pupate and hatch into imagoes.

The larvae of mealworms have a hard skin (exoskeleton) which they intermittently shed as they increase in size. Do not be alarmed at finding many of these "empty skins" in the culture, which is not evident in waxworm or maggot (fly) cultures. True worms *Anelida* spp. mature without a metamorphosis. If you see pupae in your whiteworm culture (9.8), then flies have invaded it and laid eggs on the medium, which then hatched into maggots and pupated.

Live food can be purchased in pet shops and by mail order. The production and handling is labor intensive and accordingly the prices area very high. Unless you have a dependable supplier of good quality insects whenever you need them — in my experience this has not been the case— you will find it necessary to propagate them yourself. Besides, the cost of purchasing commercially raised insects may convince you to breed your own.

While there are a number of benefits to maintaining your own insect cultures, there is a word of caution. The species of insects we can breed all year in a heated environment are also those that would happily invade your home and food storage room. You may be fond of the song of crickets on a warm summer evening out in the country, but not likely in your house.

The best place to cultivate insects is in the aviary service building or some detached out-building. With proper management the insects will stay under your control. In cooler climates they will not proliferate for most of the year outside their heated environments.

Not all insects are practical to propagate. I have propagated many food animals, including the following species of less interest to pekin robins and mesias:

- Flies *Musca* sp. and *Drosophila* sp.
- Stick insects *Carausius morosus* and *Aplopus* sp.
- Giant mealworms *Zophobia morio*
- Locusts *Locusta migratoria*
- Tropical grasshoppers *Schistocerca gregaria*
- Whiteworms *Enchytraeus albidus*
- Earthworms *Lumbricus* sp. and *Eisenia* sp.

However, I have found the following insects to be reliable and easy for on-site production, and at the same time very suitable food insects to feed to most softbill birds:

- Crickets *Acheta domesticus*
- Mealworms *Tenebrio molitor*
- Lesser mealworms (buffalo worms) *Alphitobius diaperinus*
- Waxworms *Galleria mellonella*

I have no experience with silkworms *Bombyx mori*, the larvae of a silk moth, or the "butterworm" (also called trevo, or tebo, worm in the pet food and fishing bait market), the soft-bodied larvae of the Chilean moth *Chilecomadia moorei*, which is found in the central mountains of Chile, where it feeds on the thorny trevo bush *Trevoa trinervis*. The larvae is collected and exported for fishing bait and pet food. The cost for this insect food is correspondingly high.

9.1 Insect Breeding Cabinet

The construction of an insulated, thermostatically controlled insect breeding cabinet is highly recommended to accommodate various hatching units and other types of insect cultures. The cabinet should have shelves to hold waxworm, mealworm, and cricket containers.

The temperature in the cabinet will vary from the bottom to the upper shelves, which is useful for selecting the desired temperature range for various cultures. Within reason, the warmer the insects are kept, the faster their development. Thermometers are mounted in various locations to monitor the temperature. For large-scale operations entire insect rooms or outbuildings are set up with complete climate controls.

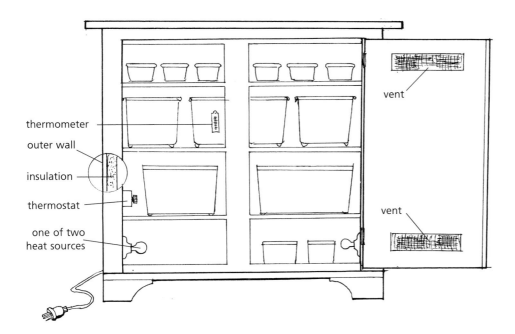

Figure 9.1-1 Insect breeding cabinet.

A retired refrigerator can be converted into a perfect insect-rearing cabinet for housing smaller containers with cricket eggs and juveniles of different development stages, in combination with the deep freezer chest if more space is required to produce large numbers of crickets.

The coolant (Freon) must be drained by a licensed tradesman, and ventilation holes need to be cut into the bottom of the door, or it can be kept ajar, to allow enough air to circulate without causing much heat loss.

9.2 Cricket Culture

The species that are commonly available in the pet industry are the house cricket *Acheta domesticus,* with a body length up to 7/8-inch (23 mm), or the Mediterranean cricket *Gryllus bimaculatus,* with a 1-3/8-inch (33 mm) body length. The adult females are distinguished from males by their 3/8-inch (10mm) long ovipositors. Theses species do not hibernate like the field crickets *Gryllus pennsylvanicus* and can thus be in a reproductive state at any time of the year. House crickets are not permitted for trade in some jurisdictions where they could become pests in homes and businesses.

Crickets can be bred in great numbers and used at various stages of their development. They have a better nutrient balance than mealworms or waxworms, so if only one of the three insects is selected, crickets would be the best choice. Crickets can also be enriched with a vitamin/mineral/amino acid supplement by dusting the insects or feeding them vitamin/mineral-enriched food in the self-feeders that are set out for the birds to catch their own food (chapter 6.10).

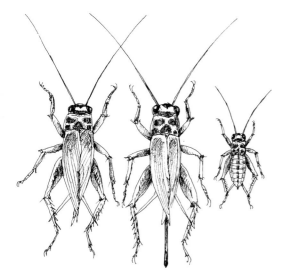

Figure 9.2-1 Adult crickets and a nymph. Male on left, female with ovipositor center, and nymph on right.

9.2.1 Cricket-rearing container

To generate and maintain a yield (or harvest) of about 500 crickets per week, they are best raised in a large rearing/breeding container approx. 24 x 32 x 16 inches high (60 x 80 x 40 cm). Crickets could be held in smooth-sided containers with walls lower than 16 inches (40 cm), but the height is needed for mounting the heating equipment and to allow space for egg cartons and other items. (Smaller hatching units (9.2.2) for the cricket eggs will also be required.)

The cricket-rearing container lid must have ventilation screens to supply fresh air. Insects have high oxygen demands, especially if the culture is well stocked. Approximately one quarter of the lid surface should be screened. The amount of air supply and heat loss can be altered by covering the opening to regulate the temperature in the unit. However stuffiness, and particularly mold formation, must be avoided. High humidity fosters the proliferation of mites

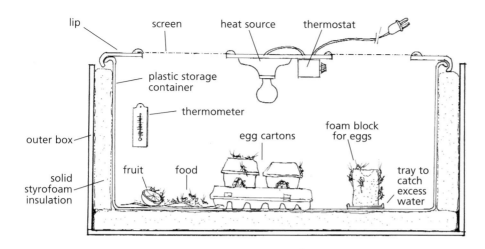

Figure 9.2.1-1 Small, insulated cricket breeder unit.

in insect cultures. Mites interfere with the hatching of insect eggs and general culture hygiene. To prevent moisture buildup the medium and container floor must be kept dry.

The temperature in the container should be about 82°F (28°C). This can be achieved by heating the space with incandescent light bulbs. The light bulb wattage is determined by trial and error to achieve the desired temperature range. To better regulate the temperature, a thermostat is patched into the electric feed and mounted under the lid of the container. There should always be a thermometer to monitor the temperature at the level where the crickets live. At least two light bulbs should be installed to back each other up, in case one burns out, to maintain some warmth.

It makes sense to insulate the unit with ¾ to 1 inch (20 mm) of Styrofoam sheeting, cased in turn by an outside box to protect it from damage and to conserve heat. This is of great benefit when power outages occur to hold the ambient temperature for longer periods. If the container is maintained in an unheated building it should be insulated to avoid heat loss.

9.2.2 Hatching unit

A hatching unit is a smaller container to receive the eggs. The unit can be a small plastic container 6 x 10 x 6 inches high (15 x 25 x 15 cm) or larger. Plastic aquariums of about that size work well. A light bulb is mounted in the lid to provide heat. Again it is practical to insulate it, as noted above, to reduce energy loss and temperature fluctuations. Insulation prevents formation of condensation on the inside walls. Freestanding water (and even drops of water) cause significant losses of newly hatched crickets as they quickly drown in it.

A hatching unit can also be placed inside an insect breeding cabinet, or the cricket-rearing container if there is room, as long as visits by larger crickets with cannibalistic notions are prevented. In that heat-sharing situation it would not be necessary to set up a special heat source for the hatching unit, nor to clad it with insulation material.

9.2.3 Starting a culture

The simplest way to get started is to purchase thirty to fifty adult crickets from a pet shop and place these in a rearing container. Three to four egg cartons are provided to offer hiding places and to capture crickets later.

A sheet of paper is used for a feeding spot. Chick starter meal, goldfish food flakes, or ground-up puppy chows are good food sources. An electric coffee grinder is ideal to pulverize puppy chow. High protein content in the food — over 25 percent — will aid growth and reproduction. Wheat bran as a sole food source for crickets is not recommended since it has a poor calcium/phosphorus ratio and a low number of nutrients. In addition, it contains phytic acid, which binds calcium and prevents its utilization.

The crickets will leave some food particles and deposit feces on the paper and elsewhere in the container, but the feeding area is easily kept clean by replacing the paper periodically. Alternatively one can omit the paper and feed the crickets on the floor of the container, but cleaning the feeding area regularly, and preventing any mold from developing, is then necessary.

Water can be supplied by constantly offering fresh fruit, lettuce, grass clippings, etc. I prefer slices of orange since these do not develop mold as quickly as other fruit. Should fruit flies discover this heaven it may be necessary to feed lettuce, dandelion leaves, etc. for a period of time in order to eliminate them. The uneaten portion of oranges, other fruit, and food mix left by the softbills, and replaced daily, are recycled to the cricket colony along with other leftover softbill food.

Small pet-drinkers may come to mind to water the crickets, but these need regular cleaning, otherwise bacteria will accumulate and cause the crickets to die off. Worse yet, the crickets crawl up inside the tubes and drown themselves in the freestanding water, creating a bonanza for unwanted bacteria. It is simpler and safer to provide water via fresh fruit and vegetables. The food can be treated with vitamins and other supplements to "gut-load" the crickets to benefit the animals to which the crickets are fed (chapter 8.13).

9.2.4 Egg-laying and incubation

Female crickets deposit their eggs with their long ovipositor into a suitable medium. Eggs are densely stacked under the surface of moist soil and in other media.

An ideal medium is a rigid synthetic foam product used for wet flower arrangements, which is sometimes offered under the name "Oasis", and is sold in blocks 3 x 4 x 10 inches long (7.5 x 10 x 25.5 cm) by nurseries and flower shops. This porous material absorbs an enormous amount of water, which is given off slowly and thus keeps the medium moist. Eggs must be kept moist at all times without reducing access of oxygen. The material is cut into smaller blocks about 3 x 3 x 4 inch (7.6 x 7.5 x 10 cm). A crater is carved into the top to make it easier to periodically add water, which spreads throughout the medium.

The moisture content can be checked by weight in lifting the block. A dry block weighs practically nothing. Depending on the evaporation rate in the environment, a foam block of the size noted may need about one ounce (28 ml) of water per day. The block can be placed on a waterproof dish to catch excess water, which will be reabsorbed by the foam. A folded paper towel can be placed under the foam block to catch excess moisture. There must not be any excess, freestanding water if the eggs have hatched, otherwise there will be a mass drowning of nymphs.

The female crickets will riddle the exposed sides of the moist foam block with egg canals to deposit a large number of eggs. When the surface is covered with fine holes, in one to three days, depending on the number of adult crickets, the block is removed and placed in the hatching unit. If the block is left in the rearing container the crickets will change their behavior and begin to dig out the eggs and eat them, tunneling and destroying the foam block in the process. This is aggravated if the crickets do not have enough moist food to stay hydrated.

The ambient temperature governs the incubation period. At 70°F (22°C) it takes about three and a half weeks and at 85°F (30°C) it takes under two weeks for the eggs to hatch. The batch hatching takes about two to four days, corresponding with the number of days the blocks were offered to the female crickets to lay their eggs. I keep the blocks in the hatching unit to water the nymphs and to provide surface to disperse. I had best results in keeping the hatching units at 83-85°F (28-30°C).

A piece of folded tissue paper is placed at one end of the hatching unit to provide hiding places. The hatching crickets (nymphs) are best fed with goldfish food flakes, but other food, as noted above, is also suitable.

For water, a slice of orange, lettuce, or a wedge of apple suffices. The hatchlings — some people call them "pinheads" — are kept in this unit until they are about ¼-inch (6 mm) long before they are transferred to the rearing container. If large numbers hatch, they should be divided up into several containers to avoid overcrowding. The small crickets will grow to adult size in about four weeks. The time frame to mature is governed by the ambient temperature and available food supply. Adults will lay eggs for

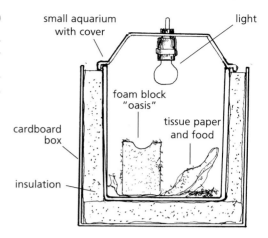

Figure 9.2.4-1 Cricket eggs laid in a floral foam block.

Figure 9.2.4-2 Creating a crater for adding water to keep the foam moist.

about two weeks and then begin to die off. Foam blocks with eggs can be stored at different temperatures to stagger hatching times ranging from 70°F (22°C) to 80°F (27°C).

9.2.5 Cleaning and handling

Once the adult crickets have completed their breeding cycle they will die off. This creates a cleanliness problem. An easy way to clean the breeding container is to shake all the crickets out of the egg cartons and sweep them and the debris to one end while placing the egg cartons at the other. The live crickets will soon be back inside the clean cartons and the debris can then be removed.

Another way to handle the lively crickets is to switch off the heat lamps to cool the container. As the temperature cools the crickets become sluggish, cling to the egg cartons, and can be transferred to a holding container while the breeding container is thoroughly cleaned with soap and water. The surfaces must be rinsed and dried with paper towels to remove soap residue.

If a lot of fruit has been fed or water from the drinkers (if you decide to use them) has been spilled, the material on the floor of the container quickly begins to develop moldy patches. It causes bad odors and becomes a source of undesirable bacterial contamination. This affects the quality of the insect food. Too much moisture creates condensation to which mites will migrate to form a dust-like film on the food and walls. As this dries intermittently, the mites die and dry onto the walls giving the crickets a foothold to climb up on the walls. It is better to keep the crickets rather on the dry side rather than too moist. Humidity can be as low as 30 percent, which inhibits mite invasions. Cricket cultures have the reputation for being smelly, but this is not the case if the environment is kept dry and the moist food is consumed before it is replenished.

To feed a portion of them, one or more egg cartons with crickets hiding inside can be placed into a plastic bag and refrigerated for fifteen to thirty minutes to slow them down. This way they are easily handled and the birds are more likely to catch them if they are fed alive. House crickets can endure low temperatures, but not freezing.

To remove large numbers of crickets, shake the egg carton contents into a tall plastic pail that is placed on the freezer floor. Then move the pail to where the crickets can be cooled down, dusted with supplement powder, and transferred to self-feeders.

Crickets of various sizes can be sorted to size by screens. Make sorting screens by cutting the bottom out of a suitable plastic container, for example, a flowerpot, and replacing it with wire mesh with different grid sizes. The crickets will try to crawl to darker places and, if small enough, escape through the screen, neatly sorted by size.

9.2.6 Large-scale propagation

There is little difference in labor output in propagating excess crickets for sale to local pet shops, and the income it provides can help defer costs for food for the bird population, plant purchases, and the various other expenses involved in aviary management.

Large-scale propagation can be done in a discarded chest freezer.[1] This is the ultimate method. These appliances are well insulated to conserve the supplied heat, plus have easy to clean, smooth walls that the crickets are unable to climb. As described above, a set of two light bulbs, controlled by a thermostat are mounted under the lid to heat the space.

The simplest way to ventilate the unit is to keep the lid raised ½ to 1 inch (12 to 24 mm). Cool air will flow in and warm air will flow out without causing condensation under the lid or along the sides. Several egg cartons are provided for hiding places. The raised section inside the freezer, which accommodates the compressor, creates an ideal shelf to place the containers with egg-laden foam blocks (hatching units). The adult crickets cannot reach this level, provided the egg cartons are not piled up too high.

It is advantageous to keep age groups separated, not only because adult crickets may cannibalize the progeny, but also to simplify periodical, thorough cleaning of the chest freezer. This can be done when the crickets are at the end of their life cycle and general die-off occurs.

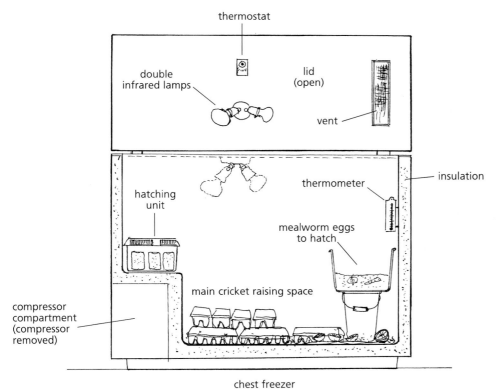

Figure 9.2.6-1 Cross section of a freezer chest, the ultimate cricket factory.

1. Theo Pagel, pers. com. 2001.

Figure 9.3-1 Wax moth viewed from above and front, larva, cocoon, and pupa.

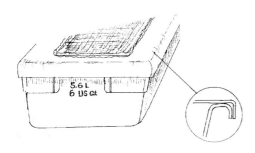

Figure 9.3.1-1 Moths lay eggs where lid and rim meet.

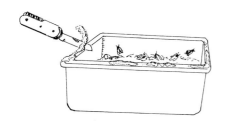

Figure 9.3.1-2 Removing strips of eggs from the container rim.

9.3 Waxworm Culture

Not really worms, waxworms are the larvae or caterpillars of the wax moth. They are a pest in beehives and can sometimes be obtained from a beekeeper when they have infested a hive.

There is a greater wax moth *Galleria mellonella* and a lesser wax moth *Achroea grisella*. Both moth species occur in the wild.

The female of the greater wax moth is up to 3/4-inch (18 mm) long, and the smaller male is 5/8-inch (15 mm) long, measured in a wings-folded position.

Their larvae grow up to about 1-1/16 inch (26 mm), just before they pupate. Food shortage will trigger premature pupation and smaller-bodied moths might hatch.

The lesser wax moth and its larvae is little over half the size of the greater wax moth.

The greater wax moth is better suited for live food production. Thirty moths of this species are sufficient to generate thousands of eggs for the next generation. Waxworm cultures are highly productive and they require little space; however, one must manage the cultures responsibly to prevent any unnecessary escapes of the insects, so as not to cause beehive and bumblebee nest infestation.

9.3.1 Breeding cycle

The cycle from egg to egg is about fifty days at 80°F (28°C), and up to three months at room temperature. By propagating starter cultures at various temperatures at the onset, we can achieve a staggered effect, to have a more continuous supply of larvae.

The greater wax moth lays up to 800 eggs and the lesser moth up to 300 eggs. The eggs are deposited in small strips between the lid and rim of the bottom part of the rearing container (9.3.4). The strips of egg masses can be scraped off with the fingernail or the back of a knife and divided up for several new cultures. Several batches of eggs can be harvested from the same container for two weeks or more. The eggs hatch within one week at 80°F (28°C) and grow to full size larvae within a month. Lower temperatures delay the development.

The larvae grow rapidly; once they reach full size they begin to wander around on the surface of the rearing container just before pupating. The fully-grown larvae like to pupate under the screen of the lid, however if too many larvae block the screen, some need to be pulled off and placed on the top of the medium or fed to the birds.

The emerging moths hatch in the rearing container and do not require special feeding, since they will begin to mate and lay eggs immediately and perish within about two to three weeks.

The pupae spin paper-like cocoons that have a small opening slot at one end. Starting (only) at this end the cocoon can be pulled apart and the brown pupae removed. Most birds relish these and feed them to their chicks. They can sense, however, if the pupae are alive and do not feed frozen, dead or dried pupae to their young.

Harvesting pupae is time consuming, but it offers important insect food that is alive, yet immobile. Pupae can be injected with medication and then fed to the birds, without the need to physically restrain them.

Fortunately the insects can be cultivated year round like mealworms. Larvae can be taken at different times depending on the size of food animal desired. By keeping maturing cultures at lower temperature the metamorphosis is delayed, which extends the harvesting and storing time.

The most practical method to start waxworms cultures is to obtain the larvae and allow them to pupate and hatch into moths to lay eggs for the next generation.

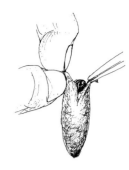

Figure 9.3.1-3 Opening the cocoon to remove the pupa.

9.3.2 Waxworm culture cabinets

Breeding colonies are best maintained in an out-building by placing the various breeding containers in a heated cabinet built for the purpose. A well-insulated cabinet can be kept in a garage or shed not connected to the home, to avoid escapees taking up residence in the house. A discarded refrigerator can be converted very nicely into an insect culture cabinet (91).

All screens must be metal since the larvae will chew through plastic screens in short order when they run out of food or approach the pupating stage. A well-populated culture will generate a surprising amount of heat on its own, hence the great need for ventilation. I found that the optimal rearing temperature is 80 – 83°F (28 – 30°C).

9.3.3 Hatching containers

Wax moths and their larvae prefer darkness and warmth. The whitish, practically hairless larvae are raised from eggs in small hatching containers. When the larvae hatch they are invisible to the naked eye and very active. Without proper food nearby, they will migrate to find it and escape from the container unless it has a tight-fitting lid and very fine metal screen. If there is always suitable food, the larvae will stay with the food source.

For the hatching container three small, round stainless-steel screens (sold for fuel funnels) can be glued into the lid of plastic storage boxes, approximately 10 x 6 x 3 inches high (25 by 15 by 7.6 cm) or larger.

A piece of paper towel can be used to line the hatching container to absorb condensation, should that occur. A ¾-inch (18 mm) layer of food medium (recipe, 9.3.3) is added with a small piece of tissue placed on top. The individual eggs within the egg strips

Figure 9.3.3-1 A hatching container for waxworm eggs.

have a diameter of <1/32 inch (1/2 mm). The egg strips are spread on the tissue paper. This allows us to monitor the hatching of the eggs. The egg cases become transparent when the minute larvae hatch and migrate to the food medium. It may take two weeks before they become evident in the culture by small patches of webbing in the medium, so be patient.

The larvae are kept in these hatching containers until they grow too large to escape through the openings of standard fly screen, at which time they are split up and transferred into rearing containers with a layer of the same food medium.

9.3.4 Rearing containers

Rearing containers are plastic storage boxes with a large metal fly-screen window in the lid, to allow for good air circulation. A useful sized container is 12 x 7 x 4 inches high (30 by 18 by 10 cm). The opening in the lid should be as large as is practical, but leaving a good size rim for gluing in the screen with a hot glue gun.

More food may have to be added to the container, depending on the number of larvae. If food is in short supply the larvae will leave the container if they find an opening.

9.3.5 Food medium

A simple and suitable food medium is prepared by volume:

- 2 parts of finely ground poultry layer pellets
- 2 parts mixed baby cereal
- 4 parts of wheat bran
- 1 part wheat germ.

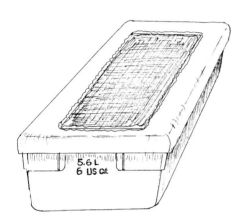

Figure 9.3.4-1 A rearing container for wax moth larvae (waxworms).

Warm up about three parts of liquid honey in a hot water bath to about 95°F (35°C) and mix the dry ingredients with the honey to produce a moist crumbly mass. It works best to use a plastic pail or large bowl and a sturdy wooden spoon to mix the ingredients. Surplus mix can be stored for a long time. A small amount of soybean protein can be added to the mix. The honey portion is varied to adjust the moisture level in the mix. Different brands of honey have different effects to obtain a moist crumbly medium.

I use liquid honey, which is usually less expensive and sold in bulk containers. This type of honey is best suited, since it does not crystallize as quickly, turning the medium into a hard, candied block. Larvae can't feed on dry, solid food mixes. To avoid this, glycerin can be added by stirring it into the heated honey at a ratio of one part glycerin to ten parts of honey. I simply increase the honey component to avoid having the mix dry up, and circumvent buying the relatively expensive glycerin. The dry ingredients can be changed in proportion.

There are many published, complex food formulas, but I found that the larvae are very forgiving as long as there is some cereal element and always honey in a moist condition. High levels of wheat bran are not recommended as explained earlier (chapter 6.10). Should the medium dry up, which would cause the larvae to starve, one can spray very little water onto the surface to recondition the medium. Too much water will cause mold, which must be avoided. If low humidity causes drying of the medium, a shallow water dish can be placed in the insect breeding cabinet to help this problem. The waxworm cultures need to be checked frequently so they do not run out of food.

9.3.6 Gut-loading

The real advantage in breeding waxworms is their readiness to consume vitamin/mineral/amino acid supplements (gut-loading), which can be stirred into the honey. For this a powdered concentrate is mixed with the honey, approximately 1:20 in volume parts for the final feeding of the larvae prior to using them as rearing food. One product, ("Prime" — a Hagen product), which turns slightly orange when it is dissolved in water, actually tinges the larvae when they consume the concentrate. This tells us that they are "gut-loaded" with the extra nutrients that we like to transfer to the birds.

Waxworms are low in calcium and high in phosphorus; the latter is plentiful in cereal, but not calcium (table 8.13). The provision of extra vitamin D and calcium is important for good bone development in the growing chicks. Without supplements, the extensive feeding of waxworms can cause, and has caused, rickets.

The injection of a concentrated cage-bird supplement solution into waxworms with a 1 cc syringe by using a 22-gauge needle is another option. Three to four larvae per chick can be injected every two days with 0.02 ml (about one drop) of the liquefied vitamin/mineral concentrate. Not all of it reaches the chicks since some of it is lost when the parents kill and soften the waxworms for feeding. This injection method can be used if the waxworms were not prepared by previously feeding the above noted enriched food and some urgency is seen to deliver supplements, for example, Vitamin D. The feeding of enriched waxworm and crickets is instrumental in the successful raising of pekin robins and other birds (chapter 6.11).

9.3.7 Handling moths

If one wishes to feed surplus moths, the container can be chilled in a refrigerator so the moths become lethargic and can then be picked out by hand or with forceps. This is also helpful when eggs are removed, to keep the moths docile. The container with breeding moths usually has some eggs in the medium, which can be saved by adding new food mix to the container. The eggs will hatch and turn into another batch of larvae in time; however, they will be of different ages, resulting in harvesting uneven-sized larvae. If some have completed the life cycle, we have to contend with escaping moths when the container is opened. It is best to work with equal-aged batches of waxworms.

Condensation on the inside of the container must be avoided. Mold will develop, which can severely inhibit the culture. This often develops when the containers are kept too long outside the warm cabinet, while picking out larvae, and when there is high relative humidity in the surroundings. The containers should always be returned to the brooding cabinet as soon as possible. As noted above, a paper towel can also be placed in the container to form a lining inside with the medium placed on it. The paper will absorb most of the condensation.

A word of caution:
Escaping larvae are a problem in a home. They will invade dark places and pupate in crevices. Usually they widen the site for pupation and chew a depression into wood, books, even plastic and other materials. Annoying moths will emerge in time as well. If the larvae (or moths) are fed to birds or reptiles they should be fed for immediate consumption or freshly killed to eliminate escapes.

9.4 Mealworm Culture

Mealworm, giant mealworm, and lesser mealworm are misnomers since they are not worms, but the larvae of beetles. These insects undergo metamorphosis during their life cycle from beetle to egg, to larva to pupa and to beetle again. Softbills consume both larvae and pupae, but not the beetles. These emit a strong scent and are not accepted for that reason. Beetles that have just emerged from the pupa shell, however, are white and soft and are accepted by birds.

Mealworms are the larvae of the darkling or grain beetle *Tenebrio molitor*. The larvae can be raised all year round because they are not dependent on seasonal rest periods like outdoor insects of northern climates, such as grasshoppers. They are the easiest food insects to manage, since they cannot escape out of even a small, glazed, 2-inch (48 mm) high dish.

Mealworms can be found in any pet shop in good quantities. They are easy to ship and store in a cool place for long periods of time. But without knowing how they were raised and cared for, they have unpredictable food value. Mealworms raised exclusively on wheat bran can be more of a problem than a blessing (11.2.4).

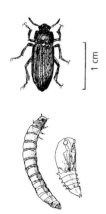

Figure 9.4-1 Beetle, larva, and pupa.

"Giant mealworms" are the larvae of the darkling or grain beetle, which have been treated with juvenile hormones (JH) to increase their natural body size. They are marketed as "giant mealworms."

The secret of raising mealworms consistently lies in preventing the beetles and larvae from consuming the eggs, keeping the hatching eggs at high humidity without creating mold, avoiding mite infestation, providing regulated temperature, and feeding the culture regularly.

9.4.1 Starting and maintaining a culture

To start a colony, I place 100 to 200 beetles in a plastic storage box about 16 x 11 x 6 inches (41 x 28 x 15 cm). They are fed with a mix of 50 percent poultry layer pellets and 50 percent wheat bran, plus sliced apples, banana peels, and carrots. The culture is kept at 80°F (28°C).

Every two to three weeks the beetles are sifted out and placed in another unit just like it, to lay more eggs. This is repeated until the beetles begin to die off. The sifted-out medium is kept at the same temperature and supplied with vegetable and fruit to provide a little moisture to the hatching medium and larvae.

By the way, the ideal place to hatch the eggs is the converted freezer chest described for raising crickets (9.2.6). Crickets generate high humidity in their air space, which is beneficial for hatching eggs. The eggs need to be kept in a covered container to protect them from being consumed by the crickets.

The larvae grow quickly at the noted temperature, but do not progress better by raising it beyond 85°F (30°C). To slow down development and store mealworms of the desired size they can be placed in cooler environments. Feeding lettuce and carrots supplies more needed vitamins than apples or oranges.

An old converted refrigerator makes an excellent mealworm-breeding unit. The various shelves can hold containers with beetles and mealworms of different size. Eggs can be hatched in this unit also, but as noted above mealworm eggs hatch well in the environment were we raise crickets, due to the relative high humidity.

A clean way to harvest mature mealworms can be done by placing folded sheets of brown paper on top of the medium. The mealworms, nearing pupation, will gather in the folds. Pupae and beetles, which have just formed are soft and whitish, as are mealworms that have just shed their old skin. Within hours the exoskeleton darkens and hardens to gain their usual color. The "white" pupae, larvae, and beetles are very good food for chicks and birds under medical care, since they are easily digested. These can only be harvested if one propagates the cultures on location and has sufficient numbers to pick out the relatively few "white" insects.

Sufficient mealworms or pupae must be retained to set up new breeding containers for beetles. When the larvae mature and approach pupation they can be observed collecting between the folds of brown paper, which is placed on top the medium. This provides them with oxygen by not being buried in the medium through the activity of younger larvae. This also reduces the risk of being consumed at the immobile stage, especially when the active larvae are starved for moisture. Placement of folded brown paper on top of the medium seems to stimulate the larvae to pupate.

9.4.2 Mite infestation

Flour mites thrive in moist and warm environments, just what we create to breed mealworms. The presence of mites can be detected by their peculiar smell and dust-like patches on moist surfaces and on moist food in the containers. Heavy infestation interferes with the hatching of mealworms. By carefully gauging the supply of moist food for the mealworms and their beetles, we can reduce the risk of mite infestation.

If need be, the humidity levels can be reduced for a period of time by airing the containers and withholding moist foods until all of it is consumed. Even then one should hold off moist food for two to three days to "dry out" the mites. The mealworms seem to be able to cope without moist food for short periods of time.

Mites can be baited with canned dog food placed into a moist paper towel. The towel and bait is removed daily and then replaced. The mites are disposed of by dipping them into hot water.

9.5 Zophobas (Superworms)

"Superworms" are the larvae of a central to tropical American beetle *Zophobas morio*. The 1½-inch (34 mm) beetle and their up to 2-inch (50 mm) long larva live on the forest floor and feed on a wide range of plant and animal food among the leaf litter. The culture medium is a mixture of slightly moistened peat moss, bark, leaf-mulch, and sand. It is kept at 76°F (25°C) with high humidity levels. The life cycle spans over five months, depending on temperature.

The larvae are rather large as live food for pekin robins, and more labor intensive to propagate, to be considered a practical culture for the average small softbill breeder. They are valuable food since they feed on dog, cat, or trout chow; but because of this and the high humidity, I have found their cultures very prone to mite infestation.

Superworms have been used to raise Hoopoe *Upupa epos*, other medium to large softbills, and reptiles.

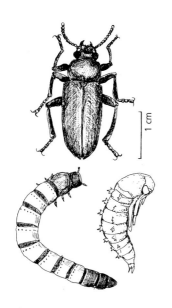

Figure 9.5-1 Beetle, larva, and pupa.

9.6 Lesser Mealworms (Buffalo Worms)

Lesser mealworms *Alphitobius diaperinus* or "buffalo worms" are the larvae of a small, shiny, black beetle about the size of a ladybug. This insect is a pest in granaries. This very prolific species can be propagated like mealworms with a shorter generation time and at lower temperatures. At 80°F (28°C) the cycle from beetle to beetle is about six weeks. The fast moving larva is up to ½-inch (12 mm) long.

Figure 9.6-1 Beetle, larva, and pupa.

This is a good food item for raising smaller bird species. Lesser mealworms have high protein content (chapter 8.13). I have found that pekin robins will accept them, but not as readily as mealworms. In fact some ignore them altogether as a food insect. Shamas *Copsychus malabaricus,* European robins *Erythacus rubecula,* tits *Parus* spp., and mesias *Leiothrix argentauris* eat them readily. The lesser mealworm feeds well on poultry layer pellets, puppy chow, trout chow, and other animal-protein based food, which makes it a valuable food insect.

The beetle flies readily and, given a chance, will infest mealworm cultures and other insect cultures. The beetles and their larvae are suspected to prey on eggs and hatching insects. The cultures must be more stringently controlled than mealworms by keeping the containers covered.

A converted refrigerator is a good propagation environment. The containers can be taller than mealworm containers, 11 x 16 x 9 inches high (28 x 41 x 23 cm), and hold about eighteen liters. They must have well-fitting, screened lids to contain the insects.

The food medium, of the same mix as described for mealworms, with the addition of ground-up puppy chow and/or trout chow, is simply put on the bottom, 3 to 4 inches deep (7.5 to 10 cm). However, it must be kept rather moist. Apples and other fruit and vegetables are used to provide moisture and a source of water. The larvae prefer apples to carrots.

Mites, as described above (9.4.2) have not been evident in cultures with high populations of lesser mealworms. Perhaps this is due the carnivorous tendency of lesser mealworms.

The simplest way to harvest them is by placing a smooth plastic dish on the medium and covering it with a folded sheet of brown paper to create a trap. The maturing larvae will wander onto the paper in search of pupating sites and fall into the tray or bowl, from which they cannot escape. This maximizes harvest ratios by removing larvae after they have grown to full size. Smaller larvae will be trapped if they run out of food and begin searching for it elsewhere.

If beetles have hatched in the culture, they too will fall into the trap. The unpalatable beetles can be separated from the larvae by simply placing all the trapped animals into a smooth sided hard

Figure 9.6-2 Screening beetles from larvae. Upper dish has 7/64" (2.6 mm) holes.

Figure 9.6-3 Trapping dish with holes or larger opening in the lid.

plastic dish, which has series of 7/64 inch (2.6 mm) holes drilled in the bottom. This screen-bottom dish is stacked on another taller one to create a dark space below. In short order the light-sensitive larvae will crawl through the holes, leaving only beetles behind. The beetles shy away from light as well, but they cannot escape through the holes. The larvae can then be screened again to remove the dust-like feces.

Another way to trap the larvae is by making a trap out of a shallow plastic food dish with a lid. Again 7/64-inch (2.6 mm) holes are drilled through the lid, food is placed inside, and the dish is positioned on top of the medium. Pieces of brown paper are leaned against it to permit the larvae to reach the lid and enter the dish through the holes. This way we collect only larvae. A larger hole cut into the lid speeds up collection, but the beetles will then have to be separated from the larvae as explained above. A piece of vitamin-enriched food can be put into the dish to gut-load the larvae. The dish must be checked daily, and emptied, since the collection of larvae will easily cause overcrowding. Too many larvae will quickly consume the food, starve, and be of lower food value.

The lesser mealworm culture produces continuously and prolifically in a non-stop mode. Every so often the entire mass is sifted to remove the thick layer of dust-like feces and debris, which accumulates at the bottom. The coarser mass is transferred to a newly set up container with fresh food. It is a good practice to set up new cultures from time to time, particularly if the beetles become too numerous. Lesser mealworms are decidedly the least demanding insects to propagate.

It is believed that lesser mealworms cause less health trouble when they are fed in larger quantity than mealworms. The relatively high protein content of greater than 24 percent could be a concern for birds with insufficient kidney and liver function, such as those with gout.

9.7 Fly Cultures

There are several species of interest to the aviculturist. The larvae, and to an extent the pupae, and the full insect are valuable food.

Fly larvae are fed by many bird breeders and are often called "pinkies". While it is a nicer term than "maggot", it can be confusing in light of still pink and hairless baby mice, which are fed to certain carnivorous birds and called pinkies as well.

Australian bird keepers propagate fly larvae extensively, and have perfected the breeding of the local bush fly *Musca vetusfissima* regularly feeding large quantities of larvae to their birds, including adult finches. The bush fly resembles the common housefly *Musca domestica*.

In essence the flies are kept in large, screened cages where they are fed sugar and given plenty of water to drink. They consume protein-rich food from a container with a maggot-raising medium that is placed in the cage, and on which the flies also deposit their eggs. The container with the food medium is left in the fly cage for two days to receive enough eggs.

The food medium consists of cereal products, milk powder and water. A bit of yeast may be helpful to initiate fermentation in a starter set. Subsequent batches of medium are split off to "inoculate" the next for fermentation. One part milk powder is mixed with ten parts of wheat bran moistened with water so that fluids ooze out when taken in hand and squeezed. The medium is kept moist and warm.

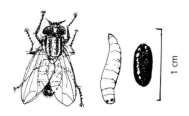

Figure 9.7-1 House fly, maggot, and pupa.

After two days the egg-collecting container is taken from the fly cage and the contents are dumped into a large bowl with more medium added. The maggots grow rapidly and can generate remarkable biomass in a short time frame. The flies eventually die off after reproduction. Holding sufficient maggots back to pupate and hatch into a new generation sets up new colonies.

I have offered commercially produced maggots or "pinkies" to pekin robins and white-eyes, but observed no enthusiasm for them, probably because I had spoiled them with other choice live food. Nutritionally they are a good food.

Maggots can be obtained from sport fishing supply stores. They should be fed to gut-load them prior to feeding them to birds and reptiles. They are usually the larvae of the green bottle fly *Lucilia* sp. or *Phaenicia* sp. I found that amphibians do not like this fly (imago) as much as the housefly, I suspect some birds may show a similar preference to houseflies.

Figure 9.7-2 Green bottle fly.

Propagating flies is of greater interest to anyone breeding fly-catchers (Other Species, chapter 12.15) and birds which hunt for flying insects; however, the flies must be prepared to be of good food value. To make the point I relay the following episode.

Case report
I bred houseflies to feed tree frogs *Hyla* spp. and experimented with boosting protein intake to stimulate egg laying. The effect was profound. Flies were fed on 2 percent milk with added protein powder for two days and released into the aqua-terrarium with several European tree frogs *Hyla arborea*. The frogs notably increased their vocalization at nightfall (to the chagrin of other members of my family) and began laying eggs for the following twenty-four hours. The frogs were then fed "empty" freshly hatched flies and egg laying stopped until several days later when gut-loaded flies were offered again, promptly followed by egg laying. Soon I had hundreds of tadpoles.

"Empty" houseflies have grey flat abdomens, which greatly extend and show yellow after the flies feed on milk. Naturally, vitamins can be added to the formula.

For small-volume fly production I used a 1-gallon (4 l), wide-mouthed jar with a screened lid. A 2-inch wide (5 cm) piece of cardboard divides the bottom into two sections. Sand is filled in on one side for larvae to pupate and food medium for the fly larvae to eat on the other. The moistened medium is made of:

- 55 g whole, coarse ground wheat
- 55 g flaked oatmeal
- 55 g alfalfa pellets or guinea pig pellets
- 1 g dried yeast
- 10 g sodium propionate (a mold inhibitor)

The lid has a 3/8-inch (9 mm) hole, which is the release opening for hatching flies later and the access port for a drinking straw to reach the medium so that water can be added if it becomes dry. The hole is plugged with cotton to prevent flies from escaping when this is not wanted.

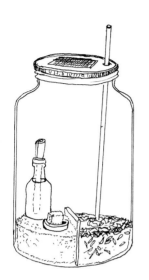

Figure 9.7-3 Small volume fly culture.

A small dish with milk powder and a sugar cube is positioned on the sand to feed the emerging flies. A small water bottle with a paper towel wick is added to provide water to the flies. Finally twenty to thirty flies are added and the culture is placed in a warm place, or into the animal enclosure. The flies eventually die or can be released after few days when sufficient eggs are laid. The growth rate of the maggots rises and falls with temperature. This method is applicable for reptiles and amphibians by placing the culture in the terrarium.

Large output and a cruder method to breed flies for birds can be done by using plastic pails with screened lids. A mixture of ground-up poultry layer pellets, wheat bran and water, plus some yeast, creates the medium for the larvae. Cottage cheese is a good addition. Flies or fly pupae (in a dish with peat moss) are added to start the culture. A piece of brown paper is a good idea to give the flies more living surface.

Another option is to fit the lid with a ¼-inch (5mm) screen to allow free flying, smaller houseflies, 1/8- to ¼-inch (3 to 6 mm) body length, to get to the medium and deposit eggs, but not the larger ½-inch (13 mm) blue bottle fly *Calliphora vomitoria*. I found that some amphibians and birds do not accept the coarse-bodied blue bottle flies.

The pails are then placed in a warm environment and left until the larvae are fully grown and travel around the surface in search for a pupating substrate. Re-feeding may be necessary. A 4-inch (10 cm) layer of peat moss–sand mixture is added. A lid with a fly

Figure 9.7-4 Large volume culture.

screen is fitted on top in which to hold the hatching flies for gut-loading.

I simply push a depression into the fine screen to hold a creamy mixture of 2 percent milk, soy protein, vitamins and mineral and baby cereal to adjust consistency to make it "hang" on the underside of the screen to feed the flies. The flies become vehicles to transfer important nutrients to the birds and their chicks. More food and water in that way is periodically added depending on consumption.

Flies are released to the enclosure by providing an escape opening, or they can be led into a clear plastic bag for freezing or cooling, to improve the catching rate of the birds. The flies will enter the plastic bag the moment the rest of the lid is covered with a darkening cloth. A number of pupae or flies must be saved for subsequent cultures.

Fruit flies *Drosophila funebris* and *D. melanogaster* are only 3-3.5 mm and 2-2.5 mm respectively in length. White-eyes *Zosterops* sp. showed interest in pursuing them for food, but not pekin robins. Fruit flies feed on yeast, certain bacteria, molds and by-products of fermenting fruit. Pet shops sell starter sets of flightless fruit flies *D. hydei* and formulas plus instructions to breed them.

9.8 Whiteworms and Earthworms

Whiteworms and earthworms belong to the phylum Annelida, the segmented worms, and the order Oligochaeta, which has about 2,400 species. They have no metamorphosis in their development and hatch from egg packages into small worms, eventually growing to adult size. Not all softbills feed on these worms. My pekin robins, mesias, and tits, for example, do not. European robins and particularly the larger thrushes feed extensively on them, and good earthworm supplies are vital for rearing their chicks.

9.8.1 Whiteworm culture

Whiteworms *Enchytraeus albidus* are also called "potworms", or enchytraeids. These are not insects, but true worms, which hatch from eggs into minute whiteworms. They grow to about 2 inches (50 mm) in length under optimal conditions.

They can be cultivated in plastic containers or wooden boxes, which are filled up to 4 inches (100 mm) with humus-rich soil and dry, weathered horse dung to create fibrous mulch. Peat moss and sphagnum moss can be added as well. The culture must be kept slightly moist at all times, but never soggy. It should be covered with a screened lid to keep insects out. Flies will otherwise lay their eggs on the food and the hatching fly larvae will compete with the less robust whiteworms for food. The cultures are kept in

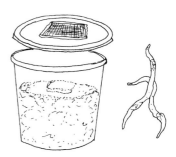

Figure 9.8-1 Whiteworm culture and enlarged worms on the right.

a dark and cool place, where they are protected from frost and temperatures rising above 80°F (28°C).

There are many recipes for their food. I have found white bread briefly soaked in water to be the best stand-by. A halved, hard-boiled egg is consumed quickly. It provides them, and the animals they are fed to, with a complete nutrient package. Chick starter and goldfish food are other good, complete foods, however, they mold more readily. The worms do not eat moldy food.

The clumps of worms can be picked away from the soil when they gather under the slice of bread. A ceramic tile can also be placed over the food to make the removal of clean worms simpler.

It is a good idea to maintain several cultures at all times, since a culture will some times "go down" and even die off, for unclear reasons. Fatty foods should be avoided. These create a pasty soil condition, which encourages mite infestation. A starter colony can sometimes be collected at a compost pile in a garden by attracting them with a moistened slice of white bread.

Whiteworms are usually not eaten by pekin robins, but by other softbills.

9.8.2 Earthworm culture

Worms of the genus *Lumbricus* and *Eisenia* are best suited for culture breeding. They can be propagated in containers with tight-closing lids, with some air holes, and approximately 12 inches (30 cm) deep. One cubic foot (30 x 30 x 30 cm) of substrate suffices to breed 100 to 500 worms, depending on species. The substrate is mixed half and half garden soil and peat moss to which sand and old leaves are added to loosen the substrate.

Figure 9.8.2-1 Earthworm egg package, hatched worms, and adult.

The worms are fed ground-up grain and corn, shredded vegetables, lettuce, fruit, grass clippings, and other soft greens. The food is placed on the substrate and covered with layers of moist sacks and a sheet of plastic to retain moisture, but never soggy. The pH should be 5.5 to 6.5 in the substrate. Food is replaced every few days when it is consumed. Temperature should be between 50 and 70°F (10 and 21°C). To obtain clean worms they are harvested by picking them from the folds of the sacks.

A large-scale production is done in a board-lined pit in a shady spot in a garden. Dimension may be 5 x 3 x 2 feet (150 x 90 x 60 cm) deep. The box is buried to half its height in the ground. A common, properly maintained compost heap provides a good environment as well.

9.8.2-2 Board-lined worm pit.

9.9 Collecting Wild Insects

Occasionally the need arises to collect wild insects. Many species of spiders can be found around buildings and gardens. Pekin robins are particularly fond of them and feed them eagerly to their chicks.

Cave crickets, field crickets, earwigs, mayflies, lacewings, smaller species of dragonflies, crane flies, moths, hairless caterpillars, termites, and grubs are all welcomed tidbits. Unfortunately, one never finds enough to depend on them. The value of "sun drenched", pollen-fed, wild insects is undisputed.

Grasshoppers, however, will occur in good numbers and body size towards late summer. I often engage neighborhood children to collect a daily supply for the pekin robin families in exchange for good pocket money.

A robin keeper soon becomes a "backyard entomologist" and insect behaviorist in discovering how to find insects and how to catch and propagate them.

9.10 UV Insect Trap

Light traps offer a very effective way to collect nocturnal insects. Removing the electrically charged screens, which kill the insect on contact, and mounting the unit on top of a ceiling fan, modifies a commercially sold UV light trap for the purpose. An angled PVC pipe connects the fan's air outflow stub to a basket with fly screen sides. The insects are forced into the container and cannot return against the airflow.

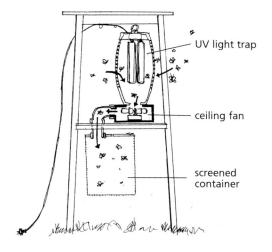

Figure 9.10-1 Home-built UV insect trap.

Depending on the season and location a hundred or more moths can be trapped nightly. Surprisingly few are damaged, which allows release of a rare specimen. I even found small tree frogs among the moths. The frogs were returned to the garden the next morning.

The moths, which have become inactive in daylight, can be carried to the food kitchen to sort the insects according to size and the birds' feeding preferences. I place the basket with door and entry port open in the aviary by late afternoon to allow the birds to pick the remaining insects and to empty the container for the next night's catch. UV light traps can be connected to the aviary or placed within it to deliver the insects directly to the birds. The disadvantage is that the yield of trapped insects is hard to establish. Many nocturnal insects have very good camouflage and are well hidden in a planted aviary before the birds begin to search for them in daylight. I found that larger moths are of particular interest to chick-rearing birds and feed them one by one to get leads as to where fledged chicks are hidden or to verify continued feeding of chicks. A softbill breeder should always attempt to get the greatest variety of insects, including wild-caught live food, for his birds.

Figure 10.0-1 A bird collection in a public aviary.

10

Conservation Breeding

Ex situ conservation breeding entails the preservation of true species and subspecies through coordinated breeding of pure genetic lines, with a high degree of genetic diversity over a long period of time. Inbreeding is avoided as much as possible within the managed population to preserve a broad range of genetic expressions, in order to make successful reintroduction of the species *in situ* possible. In essence the animals must retain their fitness to survive and reproduce in the wild, if they were to be returned.[1]

We are concerned about dimensions such as: behavior patterns, survival skills, physical and mental fitness, species-specific song, appearance (true phenotype), food and predator recognition, and others including many tangible and intangible aspects. Genetic diversity gives a population a better chance to adapt to changes by having a greater "tool box" of traits to meet a wider range of environmental conditions.

In domestic animals this is not as critical, since environments are extensively manipulated and controlled. To make the point, some of the prized breeds of canaries could not survive if they were to be released to their place of origin. It is not to belittle the achievement of intense line breeding, but it is a different branch than conservation breeding.

The zoo community uses studbooks for many species, which often includes records of the geographic origin in order to recognize subspecies. This is critical for any re-introduction program to the wild. Studbooks are based on diligent record keeping to track genetic lineage, date of hatching, individual identification, geographic origin of founders, ownership, transfers, medical history, and other data.

In any event a breeder should maintain life history records of

1. William Conway et al., pers. com. 1983.

individual birds to track breeding success, health, and behavioral aspects. The collective data of breeding events within a flock offers a valuable foundation to extrapolate trends and patterns to adjust management regimes.

10.1 Pekin Robins in Zoological Gardens

As noted in the preface, zoological gardens maintain important gene pools for endangered species; however, due to limited space and resources many species cannot be propagated in captivity. This presents an opportunity for collaboration between zoos and private breeders, with divided and important roles of exhibiting birds to the public, and off-exhibit breeding. Pekin robins and other softbills do not breed well in typical zoo environments, primarily because they are exhibited in mixed collections, which diminish the opportunity for the birds to establish an undisturbed breeding territory.

The international zoo community and related organizations work closely with the International Species Information System (ISIS). Animal inventory data are periodically submitted species by species and centrally complied on a computer system.

For the pekin robin: in April 2003, ISIS listed fifty-five zoological gardens holding eighty-six males, seventy-five females and 240 specimens of undetermined sex (401 total). Eighteen pekin robins were hatched, of which five survived, from January 1 to December 31, 2002.

This demonstrates a low rate of *ex situ* breeding, not because the zoos are not capable, but they are unable to commit space and time to breed them, in light of more urgent species recovery programs. The high number of pekin robins with undetermined sex in some institutions indicates that the establishment of breeding pairs is not a priority.

ISIS reported in November 2004: 103 males, 89 females and 195 of undetermined sex, a total of 387 birds. This population produced eleven surviving offspring for the preceding six months, the core breeding season. The reproduction rate has been 2 to 3 percent. The annual reproduction rate of birds held at my aviaries has been significantly greater in a seven-year span (chapter 6.17). Nearly all specimens were paired for breeding and given their own territory, something a zoo simply cannot commit to.

The inherent weakness of a private breeding program operated by one or two individuals is the uncertainty for continuity over the long term. Partnership with a zoological institution would be wise in the interests of preserving a valuable gene pool of endangered species.

A network of private breeders can dovetail with the exhibition of softbills by transferring birds that have contributed a high number of genetically identical offspring, and other birds not suited for breeding, to zoological gardens in exchange for birds of under-represented bloodlines.

10.2 International Studbooks

The first official register of a breeding population of animals was set up for thoroughbred horses in England in 1791. The first studbook for wild animals was proposed for the extirpated European bison (wisent) in 1923 and published in 1932.

A studbook establishes parentage and lineage of individuals, which allows for genetic management of future offspring to ensure a high degree of genetic diversity. Demographic aspects, such as geographic origin of the founders to identify subspecies status, balance of founder representation, and necessary numbers of individuals to be managed to maintain a self-sustaining population, are other benefits of studbooks.

International studbooks are established for endangered species to embrace as many specimens as possible on a global scale. The rules and procedures for the establishment and upkeep of international studbooks were first published in the International Zoo Yearbook (London Zoological Society) in 1969, and its editor was appointed international studbook keeper.

The key data includes:
- A unique number assigned to the individual
- Its house name or local record number
- Permanent ID (closed leg bands in birds, tattoo, microchip etc.)
- Sex
- Date hatched/born
- Date of death/loss
- Parentage
- History and geographic origin where possible

The World Association of Zoos and Aquariums (WAZA), the Species Survival Commission (SSC) of the World Conservation Union (IUCN), and national and regional zoo associations oversee the program.

There were over 160 international studbooks established by the year 2000 and more are added annually; beyond this there are also regional studbooks.

The International Species Information System (ISIS) developed computer software for the operation of studbooks, called Single Population Analysis Record Keeping System (SPARKS). A new program, which will replace SPARKS, called Zoological Information Management System (ZIMS), is currently under development. The American Zoo and Aquarium Association (AZA) was instrumental in creating ISIS and is its primary user.

Studbooks establish the history of a species/subspecies under human care as far back as possible, usually to the first wild-caught specimens. Data is at times verified by DNA analysis (DNA "finger printing") to confirm the genetic characteristics.

The studbook keeper and species coordinator assist in the relocation of breeding stock and progeny and thus manage endangered species on a regional to worldwide basis.

Studbooks are the foundation for cooperative breeding programs and conservation breeding.

10.3 Studbooks for Pekin Robins

Studbook keepers are sustaining a network to assist in the dispersal and acquisition of birds within the breeding consortium. The author currently maintains a regional Canadian studbook. It was established in 2000 and tracks, at the writing of this book, 187 specimens currently or formerly held in twenty-eight private facilities. Thirty-four birds are deceased birds and twenty in potentially non-breeding situations or condition.

The database assigns a unique studbook number for each bird.

Recorded are:
- Studbook number
- Left and right leg band numbers and color
- Physical, permanent features
- House name
- Sex
- Year of hatching or acquisition
- Year lost
- Reason of loss by code
- Sire and dam
- Previous and current owners
- Comments

Any development of a studbook system, be it by the zoo community or a group of private breeders, is of great value. The establishment of these data banks is urgent and meaningful in the context of species conservation.

10.4 Breeding Loans

Breeding loans are effective ways to combine potential breeding birds, which are held by different owners and by loaning birds to those who have available space. It is a well-known fact that world wide species survival plans are mostly concerned with spaces for these animals to reproduce in either their natural range (*in situ*) or in *ex situ* environments. Any available breeding space with the potential to reproduce rare animals is a precious conservation resource. Effectively, wildlife spaces are really what has become rare and what is threatened with extinction.

The matter of sharing offspring is generally handled by giving the aviary operation or institution, which is keeping and breeding the birds, half of the successfully raised, weaned offspring. The other half is divided between the owner of the male and female. If the pair has different owners, they each receive one quarter. If the breeder owns one bird he/she gains three quarters of the offspring.[1]

The ownership of offspring, if not divisible by four, lines up as follows: first the breeder, then the owner of the hen, followed by the owner of the male. The owner of the hen gets the second or third bird; the owner of the male gets every fourth bird. If the pair is breeding successfully in one season it will likely breed again in the next to even out the sharing arrangements. Breeding loan agreements and sharing of offspring should be made for multiple years.

It is understood that the breeder is making his/her best effort to maintain and breed the birds; hence possible losses are accepted as part of the course.

Shipping costs are generally born by the recipient of the birds.

The breeder bands the offspring with his/her closed leg bands and keeps accurate records of breeding success and bloodlines. The birds are registered in a studbook, where these exist.

Collaborative breeding program opportunities between public institutions and private breeders could be pursued to a greater extent for conservation breeding. Public relations and liability issues with potentially dangerous animals have understandably caused some hesitation in pursuing loan agreements. This should be further examined for softbill bird conservation breeding initiatives in the future.

---

1. David L. Bender, pers. com. 2004.

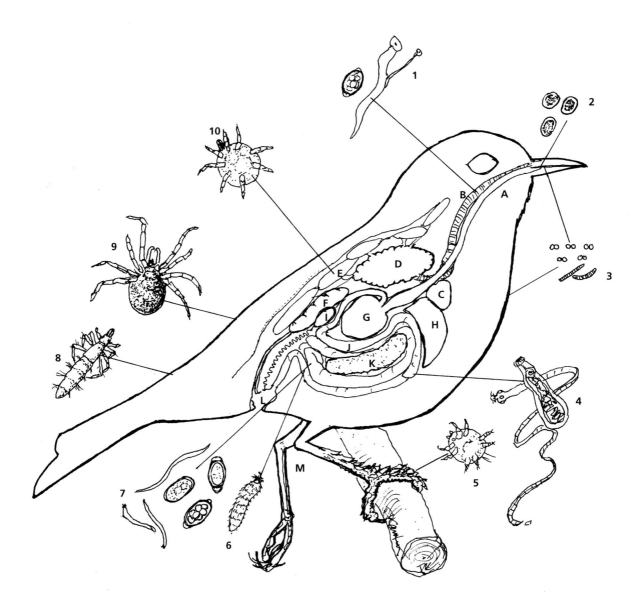

Figure 11.0-1 Sites of pathogen infections and ailments.

Pathogens
1. gapeworm plus egg
2. coccidia (oocysts, entry site)
3. bacteria and fungi (entry sites)
4. fluke and tapeworm
5. scaley leg mite
6. thorny headed worm
7. round worm and eggs
8. biting louse
9. red roost mite
10. air sac mite

Body Parts
A. esophagus
B. trachea
C. heart
D. lung
E. air sacs
F. kidney
G. stomach
H. liver
I. testis
J. instestine
K. pancreas
L. cloaca
M. leg (left leg injured)

11

Health Care

11.1 Introduction

Avian medicine is a comprehensive subject. Fortunately there are many helpful reference texts available on the general health care of birds. Very little of that information is repeated here, in the interest of brevity and in favor of referencing specific experiences in managing a breeding colony of pekin robins and other softbills. Veterinarians and other aviculturists are a good source of information on avian health care.

Good general care is the best health management program for your birds. Pekin robins and other described softbills in this book are robust and resilient birds, generally able to fend off infection.

If you have to deal with health issues it usually relates to birds you have just acquired from elsewhere. The stress of moving a bird or birds, followed by the acclimation period, depresses their immune system. The signs and behavior of sick birds was discussed under Acquisition (chapter 5.1).

Many pathogens and parasites are present in most environments, but healthy birds in good condition can usually cope with this. However, bacteria, yeast, fungi, viruses, protozoa, toxins, parasites, injury, stress, and nutritional deficiencies can cause illness; and more than one can affect a bird at a given time.

It is often rather difficult to diagnose the exact cause of illness in order to initiate targeted treatment. The most important first aid is to keep the bird's body temperature up and to reduce any form of stress. Ruffled feathers and "sitting big" indicate loss of energy and body heat. If the symptom does not change in a few hours, the bird may have to be placed in a hospital cage (chapter 3.2.2).

The location of the hospital cage is as important as the cage features. It must be placed in an environment that is free of drafts, fluctuating temperatures, and disturbances, which means one should refrain from constantly checking the patient, and exclude inquisitive children and free-roaming pets from the room.

A hen that is about to lay, or has just laid, an egg may show signs of temporary discomfort. It should be observed undisturbed for a time before it is captured and treated. This topic is explored in more detail below (11.2.6).

The problem is always how to capture a bird in a planted aviary without increasing stress (chapter 4.1 and 4.2). It is also undesirable to separate a bonded bird from its mate, because this adds to stress as well.

An infrared lamp can be set up to heat the spot where the sick bird is resting. It is not unusual to see a bird edging closer to the heat lamp, until it has reached the desired comfort zone.

This can be observed when a shipment of birds arrives after a long air transport journey. Some birds are more stressed than others, and may be soaked or soiled by the spilled water, which greatly reduces the insulation quality of the plumage and chills the birds. In addition, the birds tend to bathe to restore feather condition once released from the shipping crate. A heat lamp should always be set up, even for hardy bird species (chapter 5.8). The temperature can be allowed to reach 90°F (33°C) in one section of the hospital cage or the resting area in the aviary, as long as the birds have the choice to move to a cooler area.

Wild birds are best given this form of first aid, combined with close observation. The term "wild" bird is used in a behavioral context, i.e. wild-caught birds and their offspring, which may have been parent-raised in a spacious planted aviary where natural instincts to flee remain intact. Rest, combined with heat, is an extremely helpful treatment, with remarkable results. Often the bird can be kept in the aviary with the provision of extra heat, since we normally only have a pair or a small family in one aviary. In a flock situation we may see harassment by other birds, and certainly need to watch that the weakened bird gets enough food and rest.

If a highly contagious disease is suspected or diagnosed, the bird must be isolated as soon as possible. A sudden outbreak of such a disease is, however, unlikely if new birds are quarantined properly and access to wild bird excrements, is controlled. This is not to say that all wild birds are diseased; they are potential vectors. Unexplainable, sudden illness that affects more than one bird is alarming and a veterinarian should be consulted as soon as possible. For accurate diagnosis, time-lapse can be critical. It may be necessary to sacrifice a sick bird and perform a necropsy and various tests to isolate the problem.

Overzealous, intense handling and repeated administration of drugs under restraint ends a frightened bird's life much faster than one would expect. There is no tangible way we can relay our good intentions to a frightened bird. I frequently administer medication injected into a waxworm, which is offered from a thin stick (bamboo) to an individual bird for single or multiple treatments. This ensures that the correct amount is given to the bird in question. It may be necessary to coat the waxworm with sugar-water or honey to overcome the bitter taste of some drugs.

Treating drinking water in an aviary has limited application. It is more helpful if a bird is held in a hospital cage. Some drugs make the water unpalatable, which may cause refusal and inadequate fluid intake resulting in dehydration. Close observations are necessary to prevent this. Added sugar sometimes masks the taste sufficiently for the bird to accept it.

There are certainly effective drugs for most diseases. The difficulty lies in confirming the disease and delivering the specific drug to the bird without added stress. While it takes time to obtain results, a diagnosis can usually be made by examining fecal samples, blood, serum, tissues samples, culturing smears, x-rays, and other techniques, all done by competent veterinarians and lab technicians.

Diseases and injuries occur inevitably and, as a consolation, valuable experience is gained each time. It is helpful to maintain a log of all treatments and changes of health conditions. This information is invaluable and should be shared with the veterinarian.

11.2 Common Disorders

A few common diseases and disorders are listed here to provide only a cursory — and by no means complete — overview. Mild disorders may correct themselves with the provision of extra warmth and an undisturbed environment (using a hospital cage), vitamins, and easily digested foods. In more serious and persistent cases, a veterinarian should direct the treatment. Fecal samples should be collected and tested by a veterinary clinic.

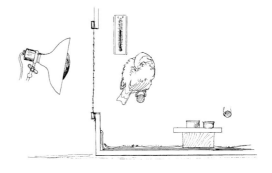

11.2-1 A bird recovering in a hospital cage with the help of infrared heat.

Specific medical treatments are not dealt with in this text, since they fall in the realm of avian veterinarian medicine rather than basic aviculture. First aid recommendations are provided.

Many drugs are prescription items; these can only be dispensed by a veterinarian, however, depending on the veterinarian–client relationship drugs can be released to the bird keeper for follow up treatment. The use of ivermectin is an example since it is a very useful drug that can be administer by the "spot-on" method (Parasites, 11.3).

See also Acclimation (chapter 5.8) to avoid health problems and losses.

11.2.1 Respiratory disorders

The bird breathes normally through its nostrils, but if more air is needed the beak is opened to increase airflow. In healthy birds this

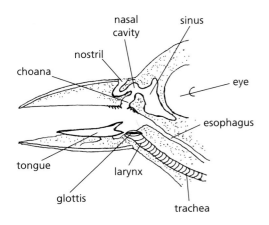

Figure 11.2.1-1 Schematic diagram of the mouth region of a bird.

is normal when they are being, or have been, chased. Some ailments cause obstructions and blockage of the nasal cavity, resulting in open-beak breathing (dyspnea).

Normal respiration rate for a bird at rest is between 60 and 70 cycles per minute, which can be counted by watching the slight tail bobbing. Fledged chicks usually show a higher rate. Monitoring changes in respiration rates is as useful as monitoring changes in weight, food intake and discharges (feces).

Signs
- labored breathing
- tail bobbing
- open beak breathing
- sneezing
- coughing
- clicking sounds with breathing
- discharges
- sitting with ruffled feathers
- lethargy
- diarrhea

Causes
- bacterial, viral, parasitic, fungal, or yeast infection
- toxic fumes and smoke
- severe stress

Teflon-coated utensils are a concern, since overheated, coated surfaces emit highly toxic fumes to birds. Birds do not belong in the kitchen for this and other reasons, specifically regarding hygiene and prevention of zoonotic diseases.

Open-beak breathing (dyspnea) can be triggered by physical exertion. Keeping the bird undistubed and removing the stimulus for fright and fear will resolve the problem. This is not an illness, but a response to an environmental condition. Prolonged exertion and anxiety will however depress the immune system and jeopardize the bird's health.

Fluid aspiration is of particular concern in hand-feeding birds. If fluid or food accidentally enters the trachea, lungs, and air sacs serious illness and death may result (chapter 7.4).

First aid
- Provide extra heat and quiet.
- Isolate in hospital cage if infectious disease is suspected.
- Increase humidity.
- Collect fecal sample and consult veterinarian if symptoms persist. (Quick x-ray investigation can indicate respiratory disorders. Other targeted tests might be needed.)

11.2.2 Gastrointestinal disorders

Signs
- sitting with ruffled feathers
- lethargy
- sleeping a lot often combined with labored breathing
- thin stool
- soiled vent feathers
- loss of appetite

Causes
- bacterial or viral infection
- parasites
- fungi
- protozoa
- poison
- severe stress
- drastic diet changes
- cancer

First aid
- Follow instructions described for respiratory disorders, above.
- Offer easily digested foods.
- Administer bismuth subsalicylate liquid. Its coating action is a good medication to treat diarrhea at its onset and in light disorders. Food items can be dipped in the medication and fed to the bird with a feeding stick in the hospital cage or administered into the beak.
- Give probiotics to strengthen and restore the bacterium flora in the gastrointestinal (GI) tract.
- Add vitamin A to diet if it is suspected to be deficient.
- Collect a stool sample to carry out fecal floatation to test for parasites and other pathogens. A clean piece of plastic foil can be placed under the roost of the bird to obtain a fresh stool sample. It should be swiftly delivered to the veterinarian lab.
- Consult a veterinarian to treat persistent disorders.

11.2.3 Sinus and eye disorders

Signs
- discharges around eyes and nares
- swelling
- rubbing and scratching sites
- sticky lids and surrounding feathers
- breathing may be affected

Causes
- respiratory infection
- injury
- vitamin A deficiency
- drafts
- cataracts

First aid
- Topical application of antibiotic eye ointment.
- Rinse with "artificial tears."
- Add vitamin A to diet (chapter 8.5).
- Improve environment, e.g. hygiene, control drafts, etc.
- Monitor for possible internal infection, which may be causing symptoms.
- Consult your veterinarian for acute treatments, which may be different than maintenance provisions.

An opaque spot in the eye may be caused by cataract especially in old birds. Consult your veterinarian.

11.2.4 Scaly legs and swollen feet

Signs
- thickened and rough scales on leg and feet due to excessive keratin build-up
- swollen pads on feet and toes (bumble foot)
- swollen foot joints and toe joints (gout)
- lameness, due to pain in joints or injuries

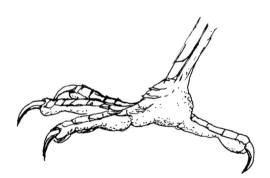

Figure 11.2.4-1 Swollen foot joint due to articular gout.

Causes
- old age combined with improper environment
  Bumble foot may be caused by improper perches that are too hard and insufficiently varied in diameter.
- mite infestation (11.3)
- vitamin A deficiency
- renal dysfunction (can cause gout)
  Uric acid crystals are deposited around foot joints causing swelling; old age, poor blood circulation and over supply of protein contribute to problems.

Figure 11.2.4-2 Infestation of scaly leg mites.

First aid
- Apply antibiotic ointment.
- Boost Vitamin A for deficiency.
- Provide branches and vegetation in the environment.
- Consistently supply fresh water.

- Take a scale scraping to be analyzed by a veterinarian.
- Check for mites. Ivermectin can be used as directed by a veterinarian to treat mites.
- Excess build up of scales and plates can be removed after softening with vegetable oil. Be sure to remove excess oil and "dry" the legs to avoid ingestion of undesirable substances.
- Controlling high protein in diets may prevent gout. Kidney failure leads to articular and visceral gout, which is usually not reversible.

Birds that have been fed consistently high protein diets may develop swollen foot joints and lameness. Mealworms are often to blame if they are over supplied and the bird becomes unable to metabolize high levels of excess protein in shedding surplus uric acid. Old birds with reduced kidney functions or birds with kidney damage are most vulnerable. Low circulation rates in foot joints lead to accumulation of excess uric acid crystals. Such swelling may take many weeks to recede, provided the diet is adjusted and the birds are kept warm to aid circulation in these sites.

If birds seem to favor a leg without visible external injury one may suspect the effects of over supply of protein in the diet.

Case report of a European robin:
A European robin arrived from an importer with swollen foot joints and persistent lameness in the right leg. It also had broken off one third of the upper mandible, which repaired itself in six months. The assumption was made that it had been on an unbalanced diet and probably too many mealworms had been fed, and thus suffered from articular gout.

Excess protein is broken down into nitrogen and converted into uric acid and excreted. An overload in the circulatory system deposits uric acid crystal in the foot joint and in time causes the visible swelling and pain. This is aggravated if the bird is dehydrated. Old age reduces the renal functions and contributes to the problem. The problem is sometimes reversible in young birds after adjusting the diet.

The European robin showed regular lameness for over three months and then gradually used its right foot again. The swelling decreased noticeably, but did not disappear entirely. More favoring of the leg and a slight increase in the swelling was seen during the cold winter months and some chronic renal dysfunction was suspected.

11.2.5 Fractures

Signs
- limbs out of position
- bird avoids using its leg or wing
- complicated breaks with penetration of the skin and bleeding
- one wing held lower than the other. If both wings are held low, another reason for discomfort should be investigated.
- bleeding caused by fracture

Cause
- flying into objects, particularly glass panes
- hanging up due to excessively long toenails
- access to a spring-type mousetrap
- predators

First aid
- isolation and quiet
- provide easy access to food and water
- trim toenails (prevention)
- secure environment (prevention)

Figure 11.2.5-1 A bird will instinctively not use an injured limb to allow it to heal.

A bird will not use an injured limb for many days or weeks and, if left undisturbed and comfortable, in most cases it will recover from the injury. A simple break will usually heal within two to three weeks.

The problem is to determine if it is a sprain or a simple fracture that still keeps the bones aligned. You may opt to take a chance if the alignment is only slightly out and the bird is getting around well by not using the limb.

Keep an eye out for harassment by other birds. Pekin robins and other social species actually gain comfort from their mate if kept together.

"Healing" may result in a permanent handicap, which may not interfere with breeding. It's your call. If the bird can be captured and examined it is best taken to a veterinarian for X-ray and treatment. Consult a veterinarian if you suspect complicated breaks.

Beak fractures can be left to heal on their own if they do not involve the live bony core near the base, and bleeding is not a concern. It may be a necessary to adjust the opposing mandible to help the bird pick up food. Sometimes we see one of the mandibles growing beyond the other. In most cases the bird corrects this by striking a hard object, and no intervention is needed.

Point of Interest:
Incidentally, behaviorists have discovered that nearly all babblers scratch their heads by bringing their foot up in front of the wing, while most other birds do it from behind the wing. This bit of trivia, however, has no significance for the husbandry of the species, except perhaps if a bird injures or breaks a wing bone, it can scratch its head without affecting the wing position.

Figure 11.2.5-2 Babblers scratch their heads from in front of the wing.

11.2.6 Head trauma

Signs
- staying on the ground
- head tilting
- moving in circles
- show paresis in a limb
- nystagmus of the eye(s)

Causes
- Often caused by hitting an object, such as a branch or window.

First aid
- Leave the bird undisturbed *in situ*.
- Provide a quiet environment.
- Provide protection from the weather.
- If moved from aviary, place in a dark, relatively cool container (hospital cage covered with a dark cloth).
- No heat lamps are recommended.
- A veterinarian may administer Dexamethasone to reduce swelling in the cranial vault.
- Provide easy access to food and water for recovery.

Prevention involves rendering clear glass opaque, placing elastic screens or fine branches in front of glass panes (chapter 3.3.3) and preventing predators from attacking and startling the birds by screen planting and hanging branches on aviary sides, and setting up night lights (chapter 4.3).

11.2.7 Egg binding

Signs
- sitting suddenly with ruffled feathers
- often sitting on the ground
- breathing heavily
- straining

Figure 11.2.7-1 A severely egg-bound hen may show extreme distress.

- showing extended abdomen and humped back
- spreading wings sideways and resting head on the ground (in severe cases)

Cause
- first laying of an egg for a young hen
- calcium and/or vitamin D deficiency
- obesity
- inflammation in the oviduct
- cold weather depleting body heat

First aid
- Do not disturb for a short while until the egg is expelled and the bird returns to normal behavior.
- Direct an infrared heat lamp towards the bird.
- If the condition prolongs, put the bird into a hospital cage with high heat and humidity.
- If the bird cannot expel the egg, contact a veterinarian.

A veterinarian or experienced bird keeper can lubricate the cloaca and massage the abdominal area to expel the egg. If that is not possible the bird may need anesthesia to break and remove the egg. This is a surgical procedure and nothing a novice should attempt to avoid egg yolk peritonitis.

Special Note: Avoid moving pairs during the active breeding season from one aviary to another during nest construction and after fledging one clutch. Hens are able to lay eggs again within days after fledging chicks, as reported under Breeding (chapter 6.6 and 6.8). Normally they would use their same restored nest or quickly build another in their familiar territory to proceed with placing eggs in a nest.

In strange surroundings, a new nest is not built in time for the hen. It causes her to lay her eggs on the ground or drop them from a branch. This unnatural condition puts stress on the bird and aggravates egg binding. Heat loss and anxiety are kept to a minimum in a nest environment, which in my estimation reduces events of egg binding.

11.2.8 Rickets and related disorders

Signs
- misshapen mandibles in young birds
- deformed leg bones in young birds, and brittle bones in adults (osteomalacia)
- stress-bands in birds growing feathers — most obvious in tail feathers

Figure 11.2.8-1 Rickets causes the mandibles fo deform.

Cause
- calcium and vitamin D_3 deficiency
- calcium : phosphorus ratio not balanced

First aid
- increase calcium and vitamin D_3 intake
- check Ca: P relationship in the diet, which should be 2:1 to 1:1.5 (chapter 8.6)
- ultra violet light (direct sunlight) helps in the conversion of pro-vitamins into highly beneficial forms of vitamin D
- provisions of vitamin supplements

Figure 11.2.8-2 Misshapen leg bones do not straighten later in life.

Deformed bones or feathers do not correct themselves. Prevention is the only cure.

11.2.9 Bleeding

Cause
- injury
- toenail clipped too short

Figure 11.2 8-3 Stress bands in a tail feather.

First aid
- Allow the bird to rest.
- Observe to ensure that the blood coagulates to stop the bleeding.
- Treat a bleeding toenail with a styptic agent, such as silver nitrate sticks.
- Remove the bird from an inappropriate cage immediately if it causes facial injuries. Repeated damage will result in considerable blood loss.
- Bleeding from the beak or vent needs the attention of a veterinarian.

Toenails with excessive length and curvature must be trimmed when necessary to avoid birds getting hung up in their environment. Try to avoid cutting into the blood vessel or "quick", unless you intend to draw blood for DNA testing (chapter 2.6). (In this case, make very small cuts each time and wait for a few seconds in between; the blood often does appear with a little delay.)

Treat a bleeding toenail while you still have the bird in hand; having to catch the bird again often aggravates the bleeding. Some literature suggests pushing the bleeding toenail into a bar of soap, or applying cornstarch with the pressure of your fingers until the bleeding stops. Application of silver nitrate sticks is recommended.

It may also be necessary to trim an uneven beak if the bird has not been able to maintain it properly. Unless it is only the tip that needs trimming, and not the bone-supported part, a veterinarian should do this repair. Uneven bills interfere with proper feeding and can lead to starvation.

11.3 Parasites

Parasites are present in most environments and live in some form of equilibrium with their host by not outright causing the host to perish, since a dead host does not allow the reproducing parasite to prosper.

Birds develop immune and defence systems to keep many parasites in check, however, the young and the stressed birds with an insufficient or compromised immune system give parasites an opportunity to grow rapidly and multiply. This often leads to secondary ailments and infections that a bird may not be able to cope with. Parasites are often species specific, hence a parasite of a finch does not necessarily affect a pekin robin or other softbill.

Most parasites produce eggs (ova), which are able to survive outside their host and can survive in the bird's environment for a long time. Some parasites have direct life cycles where the host picks up the ova or parasites and become infested, while other species have an indirect life cycle by involving an intermediate host, such as earthworms, molluscs, insects, and even mammals.

Both internal and external (endo- and ecto-) parasites can be found on birds. Good hygiene and quarantine prevents most infestations. Wild birds and rodents may be a source of parasite transmission, particularly if their feces contaminate aviaries.

New imports are a concern, especially if they come from a supplier with crowded holding facilities and frequent changes of birds. Parasite evidence can be established by microscopic investigation for species such as scaly leg mite, red roost mite, chewing lice, air sac mites and others, and usually eggs or oocysts of endo parasites such as protozoan, round worm, threadworm and gapeworm for example.

Most common signs of ecto (external) parasites
- over preening
- restless behavior
- skin or feather disorders
- skin lesions
- scaly legs

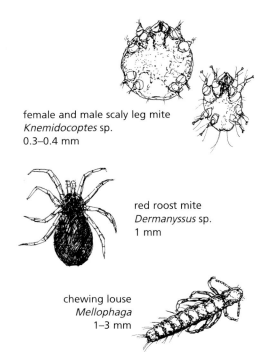

female and male scaly leg mite
Knemidocoptes sp.
0.3–0.4 mm

red roost mite
Dermanyssus sp.
1 mm

chewing louse
Mellophaga
1–3 mm

Figure 11.3-1 Ecto parasites as seen under a microscope.

Most common signs of endo (internal) parasites
- not feeding
- weight loss
- diarrhea
- increased drinking
- reduced vitality
- inability to fly off the ground
- constant sitting with ruffled feathers
- sitting with ruffled feathers
- scratching at base of beak combined with retching and headshaking
- labored breathing
- sneezing and gaping

oocysts of protozoan *Coccidia* sp.
16–25 μm

eggs of roundworm *Ascaridia* sp.
65–85 μm

eggs of threadworm *Capillaria* sp.
48–65 μm

eggs of gapeworm *Syngamus trachea*
43–46 μm

air sac mite *Cytodites* sp.
< 0.6 mm

Figure 11.3-2 Endo parasites as seen under a microscope.

The bird does not necessarily show clinical signs of being infested with parasites until parasite loads become a problem. A parasite infestation often displays bouts of symptoms interspersed with normal behavior, as the activity of the parasites within or on the host is not constant; hence the changing responses and signs of discomfort by the bird. Any abnormal behavior must concern the bird keeper. If a bird shows no improvement from labored breathing and other signs of discomfort within forty-eight hours, contact your veterinarian for a definite diagnosis.

Gapeworm *Syngamus trachea* and air sac mites *Sternostoma tracheacolum* are examples of parasites that create an off-and-on pattern of labored breathing. In contrast to this, a bird with a bacterial or fungal infection displays a more even, albeit elevated, breathing pattern.

The grey and red roost mites *Dermanyssus* spp. are more difficult to treat. These are very mobile mites; they hide in the environment and come out at night to feed on the blood of birds. It may be necessary to treat the environment to control the mites. This is not simple in planted aviaries. I have not encountered this parasite and suspect that large aviaries with only a pair of birds, plus perhaps their offspring, do not give the parasite much opportunity to find a host and proliferate.

Intestinal worms can cause weight loss and deteriorating general condition. More than one species affect songbirds. If infestation is suspected, stool samples should be collected for analysis.

I have experienced gapeworm, air sac mites, and scaly leg mites *Knemidokoptes* sp. infestations with new arrivals, but successfully treated them with ivermectin. This does not cover all bases; bacterial, viral or fungal infection, severe stress, and other causes may also be involved. It merely excludes one of the possible concerns with a relatively simple method of treatment.

Ivermectin has found wide application dealing with parasite infestation or prevention in aviculture. It has proven to be effective in eliminating a wide range of parasites. The drug works systemically and eradicates endo- and ecto-parasites feeding directly on the bird. The drug is produced for domestic, hoofed animals; for birds it must be greatly diluted. Ivermectin must be obtained trough a veterinarian who will also set up the treatment regime.

Ivermectin can be injected into mealworms and waxworms to deliver the dosage to the bird that way, or it can be added to other food items. It has a bitter taste and not all birds accept it if too much is added to a single food item. Birds need to be isolated and watched to see that they actually ingest the treated food with its measured dose.

If the bird is captured the drug can be given directly in its beak or applied to bare skin. Dabbing it onto bare skin under the wing or on the neck is called the "spot-on "method. It is a safe procedure, barring an accidental overdose. To prevent this happening, only the combined dosage for a maximum of three birds should be drawn up into the syringe for the spot-on. That way, in a worst case scenario only a triple over-dosage could be accidentally expelled onto a bird. This is usually not fatal at that level, but also no more effective then the correct amount. (An accidental soaking of the bird's skin must be avoided; even if washed off immediately the bird will likely perish within twenty-four hours.)

The treatment should be repeated in about fourteen days to eliminate a subsequent generation of parasites, which were in a protected egg stage at the time of the first treatment.

Most aviculturists use this drug routinely once or twice a year. All new arrivals from unknown flocks should be treated, unless this was done at the time of departure.

Ivermectin (10 mg/ml injectible) is also used as an environmental treatment of nest sites and cages, for example, roost mites.[1] A liter (100 cc) of stock solution is made using 2 cc ivermectin diluted in 98 cc water. As the drug is not stable in water, the solution must be used on the spot. It kills parasites on contact. If the bird is present in the environment it should not inhale uncontrolled amounts. Some inhalation is not harmful, but controlling dosage could be a problem.

At the writing of this book, the common use ivermectin has not generated profound resistance in parasites, but nature's way is to develop it over time and new drugs may need to be produced to replace ivermectin in the future. Contact your veterinarian to discuss prophylactic treatment and the use of ivermectin or other drugs that are available.

---

1. Karen Karsten, pers. com. 2006.

11.3.1 Detecting the presence of parasites

The major problem is to determine what type of parasites may be involved, if any. Clinical signs noted above are likewise indicators of other diseases (viral, bacterial, fungal infections). This makes diagnosis for targeted treatment problematic. The proof of parasites in the feces is a very helpful first step to narrow down the problem.

The detailed identification of parasites is done by trained animal health technicians, veterinarians or veterinary pathologists and thus not dealt with here. The bird keeper can, however, get some initial clues of the general type of parasites by doing simple fecal examination for his own bird population. Examination of bird droppings may reveal actual worms or segments of worms visible with the help of a magnifying glass. A microscope, however, opens a much wider window to discover adult parasites, larvae, eggs, and oocyst, protozoa, etc. A microscope with 50 to 400 magnification suffices for the cursory examination by a bird keeper. A second-hand microscope can be obtained for little cost.

A direct smear or a floatation technique is applied for the investigation.

The direct smear technique

This is a crude, but very quick method to test a stool sample.

- The dark fibrous part of a dropping, not the white substance (uric acid) is collected as fresh as possible and a small part placed on a slide mixed with a drop or two of distilled water.
- The coarse material is stroked to one side with the cover glass and the remaining fluid is then covered with a cover glass.
- The fluid is trapped under the glass and examined for parasite evidence.

We must be alarmed if a significant infestation of ova, oocysts or larvae can be found by this very basic method, and further examination and follow up is required with the assistance of a veterinary clinic.

The image at a 100 magnification is most practical to scan the slide in a line for line pattern, like reading a page in a book. A high magnification (> 200x) is necessary to identify coccidial oocysts.

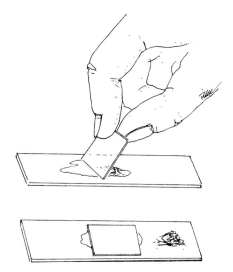

Figure 11.3.1-1 Direct smear technique.

Fecal floatation technique

This concentrates the parasite material drawn from a much larger amount of feces than the direct smear method.

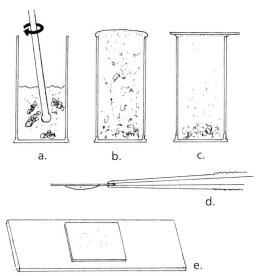

Figure 11.3.1-2 Fecal floatation technique.

a. Fresh feces is placed in a pill bottle or other small container that allow a cover glass (18 to 22 mm square) to be placed on top. (Special floatation kits can be purchased as well.)
b. Floatation solution is added to fill the container about half full and the fecal material is stirred and broken up. Lugol's iodine (5% solution) can be added to the fluid at this time to stain parasite material. Flotation solution has a specific gravity (1.200 to 1.300), which cause debris to sink and parasite material to rise to the top. The container is "overfilled" to create a meniscus (cap) and the cover glass is place on it.
c. During a ten minute waiting period the parasite material connects with the underside of the cover glass.
d. Cover glass is transferred to a slide with forceps.
e. The slide is then examined under the microscope. Commercial floatation solution, for example, zinc sulfate or sodium nitrate, can be purchased. Zinc sulfate is considered to be more effective for fecal floatation. A third option, a saturated sugar solution can easily be prepared at home by dissolving one pound of cane or beet sugar in twelve ounces of tap water and slowly heating it until it is clear.

Fecal examinations can be done routinely. If parasites are seen or suspected a stool sample from the same bird(s) should be taken to a veterinary clinic for identification and to set up treatment regimes. The keeper can team up with the veterinarian to monitor the gradual elimination of a parasite. Periodical or even daily checks can be made to see if the treatment is affecting a parasite count.

All new bird acquisitions should be tested this way when they arrive and before they are released from the quarantine unit. Ecto-parasites (mites and lice) can be found if the bird is handled for banding, weighing, crating, toenails clipping, etc. by handling it over an open plastic bag and examining the debris that collects with a magnifying glass or microscope. Veterinarians also perform skin scrapings to search for evidence of burrowing mites.

Digital cameras work quite well in capturing the image of parasites under a microscope for electronic communications with the animal clinic and lab.

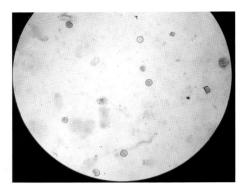

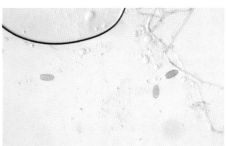

Figure 11.3.1-3 Oocysts of a coccidium (top) and eggs of a roundworm *Capillaria* sp. (bottom).

11.4 Exertion Myopathy (muscle burn out)

This condition is addressed again here (see Breeding, chapter 6.12) because it occurs all too often with breeding pekin robins and related bird species. It is caused by overexertion of the muscular system due to repeated and unsuccessful attempts to fly up on a perch for a safe retreat or to join up with siblings. Extended periods in a flight reflex mode cause a metabolic disorder affecting the bird's mobility.

Extensive harassment by aggressive cohabiting birds, pets and predators can result in this form of stress. Left untreated, it can lead to death.

Signs
- mainly seen in fledged chicks
- extreme weakness
- sleeping on the floor
- inability to perch
- uncoordinated moves
- paralysis

Cause
- extreme physical exertion
- severe stress and shock

First aid
- Transfer to hospital cage.
- Provide high temperature, 90°F (32°C).
- Maintain absolute quiet.
- Force feed easily digested food dipped in electrolytes until the bird feeds itself again.

Recovery is slow. It can take up to a week to regain muscle control.

Prevention
- Provide ample vegetation cover, "laddered" branches, and perches in the aviary when chicks are fledging.
- Block brightly lit cage fronts or sides, which might cause chicks to move towards the light, and/or direct artificial light towards dense vegetation.
- Respond to the natural instinct to perch in a secluded place.
- Resolve conflict among birds in the aviary.
- Reduce shipping stress.

This health problem can be avoided. Inappropriate environments and too much disturbance are the cause. Breeders often report losses of chicks between the second and fourth day after fledging, which mirrors my experience prior to discovering the root of the cause and adjusting the aviary environment (chapter 6.12).

Figure 11.4-1 Conifer branches provide a ladder for chicks to reach high roosting places.

11.5 Stress

Stress has become a household word for physiological or psychological events of discomfort. Stress in a clinical sense is expressed in changes of body functions, blood chemistry, tissue lesions (stomach ulcers), behavior problems and other effects. Stress weakens the immune system and vitality of a bird to a point where it could be lethal in itself, however more typically it opens the door for other diseases. Stress is caused by stressors, which constitute an almost endless list of undesirable situations and environmental conditions in the life of a bird. Stress, as such, is difficult to diagnose. This is where the sixth sense or intuition of a bird keeper is critical.

Signs
- depression
- lethargy
- throwing eggs or chicks out of the nest
- hyper-aggression
- hyperactivity
- stereotyped movements
- aberrant behavior
- indigestion
- open-beak breathing (dyspnea)

Causes
- fear
- inappropriate diets
- overcrowding
- wrong photo-periods
- extreme ambient temperatures
- wet and drafty environments
- inter- and infra-specific aggression
- territorial conflicts
- parasite loads
- over exertion (myopathy)
- shipping
- capture and handling
- predators and rodents
- lack of hygiene (bacteria, protozoa)
- toxins
- excessive noise
- medication

First aid
- removal of stressors
- correction of improper husbandry and environments

- disease control
- increase observation
- monitor activity (security camera)
- review and or make relevant diary entries

11.6 Starvation

Starvation is a serious issue. An accidental lack of food and water results in many losses in as short a time as twenty-four hours.

Probably more softbills are lost through this human error than through disease. This is a "management disease" that we must be conscious of and constantly strive to prevent by setting up and following proper routines and safe guards.

Monitoring and food presentation to ensure intake, and bullying by dominant birds is noted in Feeding (chapter 8).

11.7 Weather exposure

Extreme high and low temperatures are causes of death. Common sense, and careful observation and monitoring of the bird population are the best approach. Draft and driving rain are a much greater problem than still, cold air in the birds' environment.

FINAL NOTE

While you may be intimidated and overwhelmed by the bewildering array of diseases, in practice you may seldom have to deal with them. Unless you regularly import birds to your aviaries, or ignore good husbandry methods, you may find that oversights and mistakes in daily care are more of a threat to birds than infectious disease.

As stated frequently, maintaining diaries and records are invaluable tools in good aviculture. I keep my diary on the night table and make entries before I turn out the lights for the day.

Figure 12.0-1 Softbill birds embrace many orders and families of bird species. Note: Key to bird species on page 199.

Other Species of Softbills

12.1 Introduction

The species presented in this chapter were selected for the similarity in their care and breeding to the pekin robin. This is not a comprehensive account of suitable smaller species for aviculture, but a cross section of some relevant families and species that have been available in aviculture in the past.

While I consider the management information for pekin robins provided in this book to be a good basis for venturing into caring for the genus *Leiothrix* (pekin robins and mesias), I suggest consulting pertinent literature and getting input from experienced aviculturists for the other species introduced in this chapter. The inspiration to examine other small species of softbills is spawned by the appeal and diversity of this beautiful group of birds, but also by the awareness that more work is needed to breed them more consistently in our aviaries.

Some of the species noted here can be housed with pekin robins in a community aviary, but successful breeding becomes more difficult in mixed-species enclosures. The size and features of the aviary have a critical influence on breeding success. Beyond this, individual traits of the birds play a role. Each aviary, and its inhabitants, is unique and what works in one cannot necessarily be achieved in another. In the end, only trial and error will determine what works.

Several food stations and regular food supply diminishes the drive to defend feeding and nesting territories. As advocated previously, environmental enrichment — ample vegetation and other habitat "furniture" — helps to shrink the size of breeding territories by creating sight barriers and a sense of privacy.

As a general rule, the closer a species is related to another the more conflict can be expected. Besides direct intra-specific aggression, there are indirect conflicts as well. For example, a pair of pekin robins dismantled a chaffinch *Fringilla coelebs* nest-in-progress in search of spiders when the birds saw the cobwebs that the hen chaffinch had used to anchor the nest to a branch; and a small flock of goldfinches *Carduelis carduelis* defoliated a stand of bamboo, which had concealed a nest site of pekin robins.

Naturally, larger or aggressive species with carnivorous tendencies, which might become nest robbers, should not even be contemplated. The temptation to add another species to relatively large softbill breeding aviaries is often hard to resist. More finesse to create potential breeding territories and closer observation to monitor interaction of all birds can lead to success. For most species the most reliable approach is by giving each breeding pair of a territory of their own.

The availability of softbills changes with import and export restrictions, which are periodically imposed by national and international regulatory agencies to curb the spreading of infectious diseases to domestic stock or humans and to preserve wild animal populations from becoming endangered and extinct. Many species are regularly bred in aviaries and available to the softbill enthusiast. Avicultural magazines and journals often list available, aviary-bred birds, and specialized importers offer birds for acquisition.

Zoological publications list species of animals in taxonomic order based on the widely accepted concept of evolution by the English naturalist Charles Darwin (1809-82). Bird books reflect that sequence by beginning with the most ancient order the flightless Kiwis and progressing to the most modern order of perching birds. The alignment of the families within the order is not always consistent between publications and neither are genus and species names. The taxonomy used in this text will be that of Sibley and Monroe's *Distribution and Taxonomy of Birds of the World* (1990) and its 1993 Supplement. Common names vary greatly.

This chapter addresses softbills, which belong to the order of Passeriformes, the perching birds. Beyond it, there are many others passerine and non-passerine orders and families deemed to be softbills. Several are illustrated in the composite drawing 12.0-1. The key to this illustration of various species is given on page 199.

The birds in the following pages are introduced in order of similarity to managing and breeding pekin robins and not strictly by taxonomy. Some are popular aviary birds, which have been rarely bred in the past. Hopefully this book will add to the initiative and inspiration to advance aviary breeding to create self-sustaining populations for aviculture, and conservation programs for a wide range of softbills.

I have managed most species in one way or another in a zoological garden or my aviaries, but I have not bred all of them. Personal communication with other breeders and available literature fill in the experience gaps. The short paragraphs and illustrations may serve as a starting point to explore the keeping of other species, and serve to compare notes among fellow bird keepers.

12.2 A Palette of Other Passerine Softbills

I cannot resist the term "palette" when I look at the marvelous spectrum of colors in the softbill species that are listed and illustrated in this chapter. Due to the layout of the paintings, the species are not necessarily drawn to scale to each other.

Order: PASSSERIFORMES

Family: SYLVIIDAE
Tribe: Timaliini (babblers)

silver-eared mesia	*Leiothrix argentauris*
red silver-eared mesia	*Leiothrix argentauris laurinae*
blue-winged minla	*Minla cyanouroptera*
red-tailed minla	*Minla ignotincta*
chestnut-tailed minla	*Minla strigula*
whiskered yuhina	*Yuhina flavicolis*
striated yuhina	*Yuhina castaniceps*
black-chinned yuhina	*Yuhina nigrimenta*
white-crested jay thrush	*Garrulax leucolophus*
greater necklaced laughing thrush	*Garrulax pectoralis*
yellow-throated laughing thrush	*Garrulax galbanus*
spectacled jay thrush (hwamei)	*Garrulax canorus*
bearded tit	*Panurus biarmicus*

Family: AEGITHALIDAE (long-tailed tits)

northeastern long-tailed tit	*Aegithalos caudatus caudatus*
central European long-tailed tit	*Aegithalos caudatus europaeus*

Family: PARIDAE (true tits)

azure tit	*Parus cyanus*
blue tit	*Parus caeruleus*
coal tit	*Parus ater*
varied tit	*Parus varius*

Family: PYCNONOTIDAE (Bulbuls)

white-cheeked bulbul	*Pycnonotus leucogenus*
red-eared bulbul	*Pycnonotus jocosus*
red-vented bulbul	*Pycnonotus cafer*
black-crested bulbul	*Pycnonotus melanicterus*

Family: ZOSTEROPIDAE (White-eyes)

Oriental white-eye	*Zosterops palpebrosa*
Japanese white-eye	*Zosterops japonica*
chestnut-flanked white-eye	*Zosterops erythropleura*

Family: MUSICAPIDAE (Thrushes, Flycatchers, etc.)

white-throated thrush	*Zoothera citrina cyanotus*
dama	*Zoothera citrina*
European song thrush	*Turdus philomelos*
European black bird	*Turdus merula*
Siberian thrush	*Zoothera sibirica*
redstart	*Phoenicurus phoenicurus*
white-spotted bluethroat	*Luscinia svecica cyanecula*
red-spotted bluethroat	*Luscinia svecica svecica*
Siberian rubythroat	*Luscinia calliope*
European robin	*Erithacus rubecula*
shama	*Copsychus malabaricus*
Oriental magpie robin or dayal	*Copsychus saularis*
Asian paradise flycatcher	*Terpsiphone paradisi*
verditer flycatcher	*Musicapa thalassina*
Japanese blue flycatcher	*Cyanoptila cyanomelaena*
rufous-bellied niltava	*Niltava sundara*

Family: IRENIDAE (Leafbirds)

Asian fairy bluebird	*Irena puella*
greater green leafbird	*Chloropsis sonnerati*
golden-fronted leafbird	*Chloropsis aurifrons*
orange-bellied leafbird	*Chloropsis hardwickii*

Family: FRIGILLIDAE (Finches, Tanagers, Honeycreepers, etc.)

blue-winged mountain tanager	*Anisognathus flavinuchus*
superb tanager	*Tangara fastuosa*
Brasilian tanager	*Ramphocelus bresilius*
blue-naped chlorophonia	*Chlorophonia cyanea*
violaceus euphonia	*Euphonia violaceus*
red-legged honeycreeper	*Cyanerpes cyaneus*
purple honeycreeper	*Cyanerpes caeruleus*
green honeycreeper	*Chlorphanes spiza*

Chapter 12 ■ OTHER SPECIES OF SOFTBILLS

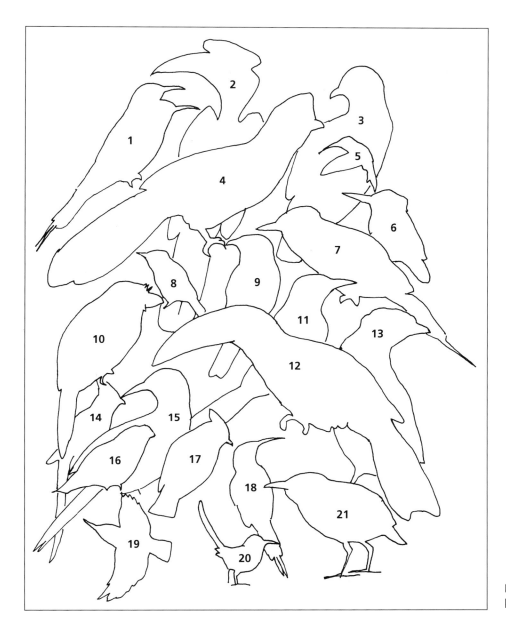

Key to figure 12.0-1
[Birds are not to scale]

1. European bee-eaters *Merops apiaster*
2. giant hornbill *Buceros bicornis*
3. Sulawesi green imperial pigeon *Ducula aenea*
4. red-crested touraco *Tauraco erythrolophus*
5. sparkling violet ear hummingbird *Colibri coruscans*
6. grey-headed kingfisher *Halcyon leucocephala*
7. lilac-breasted roller *Coracias caudata*
8. blackcap warbler *Sylvia atricapilla*
9. red-backed shrike *Lanius collurio*
10. fire-tuffted barbet *Psilopogon pyrolophus*
11. green jay *Cyanocorax yncas*
12. red-billed toucan *Ramphastos tucanus*
13. hill mynah *Gracula religiosa*
14. blue-naped mouse bird *Urocolius macrourus*
15. golden-breasted starling *Cosmopsarus regius*
16. green broadbill *Calyptomena viridis*
17. Bohemian waxwing *Bombycilla garrulus*
18. orange-breasted sunbird *Nectarinia violacea*
19. velvet-fronted nuthatch *Sitta frontalis*
20. blue and white wren *Malurus leucopterus*
21. banded pitta *Pitta guajana*

12.3 Silver-eared Mesia
Leiothrix argentauris

Color plate 12.3-1
TOP: a pair of silver-eared mesias with a fledged chick.
BOTTOM: a pair of red silver-eared mesias, male in foreground.

The silver-eared mesia is closely related to the pekin robin and is of the same genus. Their range is similar to that of the pekin robin but extends further south to Malaysia and Indonesia. Mesias were commonly imported in the past but, like the pekin robin, they are now listed on Appendix II of CITES.

There are regional subspecies with some color variation and slight difference in body size. The red silver-eared mesia *L. a. laurinae* from Sumatra is more colorful and larger than the northern subspecies.

Mesias are dimorphic and easy to sex as adults; the males are more colorful than the hens. Juvenile plumage is similar for males and females except the upper and under tail coverts are more reddish brown in males.

Their calls are louder and less melodic than that of the pekin robin. Male-chicks have a loud, penetrating contact call like the adult male, not given by hen-chicks or adult hens.

They are hardy in outdoor aviaries with winter protection. I found that they choose outside sleeping roosts despite the availability of warmer shelters. As a precaution, the birds are locked into protected shelters when the temperature drops two to three degrees below freezing.

Silver-eared mesias tend to pull leaves off bamboo plants and defoliate upper branches, particularly in the winter season. Conifers (cedar and fir) are a good alternative for their aviary.

Size: Silver-eared mesia have a body length of about 7 inches (18 cm). Pekin robins in comparison are about 6 inches (15 cm) long.

Diet: Same as pekin robins but more insects should be offered. Mesias readily eat lesser mealworms (buffalo worms). A variety of insects are needed.

Breeding: Mesias can be bred following essentially the same management regimes as for pekin robins. Breeding pairs, and males in particular, seem to be less tolerant of disturbances. Scolding and warning calls are emitted more frequently and extensively than with pekin robins. This can interfere with the nesting of pekin robins close by. (Silver-eared mesias and pekin robins can be housed together outside their breeding season, although silver-eared mesias may be more aggressive and dominate feeding stations.)

In my experience, the adults seem to be more nervous when nesting, and prone to abandon the chicks due to disturbance. Males become more agitated if something is out of order and are the first to stop feeding the chicks. While the hen may continue to feed, his behavior influences the hen and she will abandon the nestlings within a day or two as well.

Silver-eared mesia chicks have sturdier leg bones than pekin robins and waiting past day seven with closed banding can be problematic. Banding sooner leaves the chicks with insufficient feather-covering to hide the band. Even then the adults may detect the leg band and make determined attempts to remove it from the nest — chick attached. A security camera reveals that the adults are constantly clearing the nest of any undesirable objects. I remove the chicks for hand rearing if the parents neglect them after banding.

Hand rearing can be done with good success once the chicks are over five days old. Live insects must be of good quality and variety while chicks are reared by the parents or by hand.

12.4 Minlas

blue-winged minla	*Minla cyanouroptera*
red-tailed minla	*Minla ignotincta*
chestnut-tailed minla	*Minla strigula*

■ Blue-winged minla

Also called a blue winged siva, this bird has a similar range to the pekin robin, but with greater southward extension. The sexes look alike. In good plumage and kept in the same environment, one can see a slight difference in the facial markings of the male. The white band above the eye is of a brighter white and more distinct than in the hen. Blue-winged minlas are very active and should not be kept in cages, but in planted aviaries. Most of the available birds are wild-caught.

Size: Close to pekin robin size, 6-¼ inches (16 cm), but of slimmer shape.

Diet: Same as the pekin robin but with more insects.

Breeding: The breeding and care regimes are the same as with pekin robins. These birds tend to be more secretive and hide their nest well. They are very active and less trusting than pekin robins and should have larger and more undisturbed aviaries to stimulate nesting.

■ Red-tailed minla

Another babbler, which has a range (Nepal to SW China) close to that of pekin robins. It can be found in mountain forests regions up to 11,000 feet (3,400 m) elevation during the breeding season, and at lower elevations in the winter. Minlas often travel in small flocks with pekin robins in the winter season when they migrate to lower elevations.

The male shows more striking orange-red edges on the tail feathers and primaries, which are pale yellow on the hen. These are very spirited birds, highly active, most inquisitive, and rather tame by nature. They remind me of small species of tits *Parus* spp. in their behavior.

Size: 5-½ inches (14 cm) body length.

Diet: The same as the pekin robin, but with more insects.

Breeding: Breeding and care regimes are the same as for pekin robins; clutch size, incubation, and nesting period are the same. They are not easily intimidated and nested with a pair of pekin robins in a planted aviary measuring 7 x 8 x 9 foot high (2.1 x 2.4 x 2.7 m). They built their nest in an open nest box, which had been intended for orange-headed ground thrushes, and used various plant fibers and rootlets to create the base, but finished the nest cup with coconut fiber.

While the species is rather easy-going and tolerant to other, even larger, birds in their breeding aviaries, they do best as single pairs in a dedicated aviary. I found that their energy and curiosity interfered with larger species trying to nest.

■ Chestnut-tailed minla

is of the same size as the red-tailed minla. Sexes are alike in plumage color. The species is cared for in the same manner as the above species.

Color plate 12.4-1
TOP: a pair of blue-winged minlas, male in front.
CENTER: red-tailed minlas, male in center.
BOTTOM: a chestnut-tailed minla.

12.5 Yuhinas

whiskered yuhina	*Yuhina flavicollis*
striated yuhina	*Yuhina castaniceps*
black-chinned yuhina	*Yuhina nigrimenta*

Color plate 12.5-1
TOP: whiskered yuhina.
CENTER LEFT: striated yuhina.
BOTTOM RIGHT: black-chinned yuhina.

Various species of yuhinas can be found in aviculture. They are Asian babblers as are the aforementioned species, but are more delicate to acclimate and to care for.

Yuhinas are found in southeastern Asia from the Himalayan Mountains southward to Thailand and Laos. Some species breed at elevations as high as 6000 feet (1800 m). Yuhinas form small flocks of mixed species and migrate to lower elevations in the winter. They are very social, busy little birds, similar to chickadees in their behavior. They are not great singers, but frequently utter contact calls like other babblers.

Yuhinas should be kept in pairs and not as single birds. They are a contact species that roost together and groom each other as described in the pekin robin. Yuhinas are fairly hardy in out-door aviaries, but need more winter protection.

Size: whiskered yuhina: 5-½ inches (14 cm)
striated yuhina: 5-½ inches (14 cm)
black-chinned yuhina: 4-½ inches (11.4 cm)

Diet: The same as the pekin robin, but greater amounts of insects and a nectar drink.

Breeding: Sexes are alike in appearance. As with other monomorphic species it is best to acquire several birds to have some assurance to find a pair among them, because importers are usually not set up to carry out DNA testing for these small birds.

Yuhinas are not commonly bred in aviaries. A habitat enclosure with dense vegetation extending to the in the lower regions, with tree stumps and ground cover offers the best chance to breed them. Some species are rather territorial (whiskered yuhina) and will not accept another pair in their aviary during breeding season, while others demonstrate cooperative nesting behavior (striated and black-chinned yuhina). The dominant pair is the most likely to nest successfully.

Nesting material should include mosses, hair spider webs and fine plant fiber in addition to coconut fibers.

12.6 Laughing Thrushes

white-crested laughing thrush	*Garrulax leucolophus*
greater necklaced laughing thrush	*Garrulax pectoralis*
yellow-throated laughing thrush	*Garrulax galbanus*
spectacled jay thrush or hwamei	*Garrulax canorus*

The close to fifty species of laughing thrushes belong to the babbler tribe (Timaliini), genus *Garrulax*. They earned their name due to their noisy contact calls as they travel through the underbrush and treetops. They are very active birds that explore their environment including the forest floor and anything that may harbor an insect, small vertebrate animals, fruit, and vegetable matter.

They occur from India to the Himalayas southward to Borneo. Some species migrate to high elevations in the mountain forests to 16,000 feet (4,800 m). They are relatively hardy and can be cared for much like pekin robins, with perhaps more space.

Laughing thrushes, particularly the larger more pugnacious species, should not be kept with smaller birds, because they may eat the eggs and nestlings and may hunt down fledglings. Even outside the breeding season, they must be introduced with caution to a community aviary to monitor interrelationships with other birds (stress, aggression, competition for food, etc.).

The spectacled jay thrush — also called melodious jay thrush or hwamei — is a rather plain looking bird, but its song is superb and much coveted.

Size: white-crested laughing thrush: 11 inches (28 cm)
greater necklaced laughing thrush: 13 inches (33 cm)
yellow-throated laughing thrush: 9 inches (23 cm)
spectacled jay thrush: 10 inches (25 cm)

Diet: The same as for pekin robins, with more fruit and hard-boiled egg; baby mice and zophobas ("superworms").

Breeding: The sexes are monomorphic. These active birds need large habitat aviaries for breeding. Single pairs should be kept per aviary and maturing young must be removed because the parents may become aggressive towards them.

The cup-shaped nests are built with plant fibers, moss, rootlets, dry leaves, and other plant material. Laughing thrushes will nest in half open boxes and accept nest supports like pekin robins. Stands of bamboo and conifers offer good nest sites.

They lay four to five eggs with an incubation period of thirteen to fourteen days. Both parents participate in the incubation.

The chicks are fed live insects. Laughing thrushes are rather carnivorous and may become frustrated if live insect supplies are not varied and plentiful, which can lead to killing and consuming their own chicks. Moving the feeding station and hiding food can help keeping the birds focused on hunting for food and feeding their chicks.

Color plate 12.6-1
TOP TO BOTTOM:
white-crested jay thrush,
greater necklaced laughing thrush,
yellow-throated laughing thrush,
and spectacled jay thrush.

12.7 Bearded Tit or Reedling

bearded tit   *Panurus biarmicus*

Color plate 12.7-1
TOP RIGHT: juvenile male.
CENTER: male.
BOTTOM LEFT TO RIGHT: female and juvenile female bearded tit.

The bearded tit is grouped with the babbler family. It is the only European representative of the tribe Timaliini. It ranges throughout Europe, however in pocketed populations with a more contiguous distribution in the region of the Black and Caspian Seas and eastward to Mongolia.

They have longish tails and the male is distinguished by the black moustache-like stripe, black under tail-coverts and a bright yellow beak. The hen has beige-brownish colored beak. Young have beaks that are more yellow than juvenile hens. The male has an orange iris, which is green in the hen and grey in the juveniles.

This is a delightful aviary bird. The mates have a close relationship, groom each other and search for food by remaining in close contact, which is typical behavior for this family. The song is a simple, quiet twittering.

Size: The bird is about 6-½ inches (17 cm) long; the tail length is just over 3 inches (8 cm).

Diet: The species is insectivorous during the nesting season, but converts to consuming seeds in the winter, similar to other tits. But insects should be offered throughout the year, in addition to a softbill diet.

Breeding: Bearded tits are often bred in biotope-type aviaries. As their name implies, their preferred habitat is stands of reeds and cattails. This should be simulated in the aviary to encourage breeding, general wellbeing, and nesting.

Bearded tits nest in small, loose colonies in the wild, so more than one pair can be kept in an aviary. Single adults may even participate in the caring of chicks of breeding pairs.

Mated pairs stay together for life. Courtship involves a form of chase by flying repeatedly upwards and tumbling through the reeds. The male lowers his head, fans his tail and sings while bowing to the hen.

One or more bundles of dry cattails and reeds should be provided to them. Medium size and dense species of bamboo serve as a substitute for reed and cattail patches. The male selects the nest site, often just above the water surface or wet, marshy ground. Both work on building the nest. The cup-shaped nest is made of plant material taken from the surroundings. The nest is lined with woolly plant fiber and feathers.

Four to seven eggs are laid in a cup-shaped nest and two broods may be raised per year. Both parents incubate. Incubation period is eleven to thirteen days; both parents brood the chicks. They fledge when they are about two weeks old.

The young widely disperse after fledging at first, but congregate into flocks of young birds from several nests as they become independent.

12.8 Long-tailed Tits

northeastern long-tailed tit *Aegithalos caudatus caudatus*
central European long-tailed tit *Aegithalos c. europaeus*

Long-tailed tits belong to the family Aegithalidae, which have very short bills and tails longer than half their body length. They range throughout Europe to Asia Minor, Russia, and Japan in several subspecies, where they live predominantly in broad leaved-mixed forests.

Tits actually change their stomach lining for the winter season to enable them to live on seeds and nuts, and revert back to an insectivore's digestive system in the spring. A late, extended cold snap in the spring, which drastically reduces availability of insects, becomes a serious challenge to tits and many may be lost to starvation.

The birds sleep in tight clusters to stay warm during cold weather. If only one pair or a single bird is held in an aviary it is necessary to provide shelter from severe weather in lieu of the warmth of a group of birds sleeping together. Sleeping boxes must be provided for the winter.

Size: 5-½ inches (14 cm), weight 5 to 6 gm

Diet: Tits feed primarily on insects during the summer and add seeds to their diet in the winter. Insectivore mix, hazel nuts, pine nuts, sunflower seed, suet, etc. are provided to them in aviaries.

Mealworms and their pupae, waxworms/pupae/moths, buffalo worms and pupae, plus other insects are eaten. Crickets are not always accepted. Aphids and insect eggs are part of their natural diet.

Breeding: Long-tailed tits can bred in a habitat aviary. Adequate nesting sites and nesting material must be offered. Cobwebs should be collected for them. Gather cobwebs by raking them onto fine twigs and tying these to branches in the aviary. I have found that offering cobwebs literally triggers nest building in other species (chaffinches).

These birds build an elastic, roomy, egg-shaped hanging nest made of cobwebs, moss, and lichen. The inside may be lined with a thousand and more feathers. The nest is about 7 inches (18 cm) long and resembles the longer nest of the related North American bushtit *Psaltriparus minimus*. It is suspended from branches and has an entrance hole near the top. The intricate nest can take two weeks to be completed. These birds are colonial nesters and more than one pair may use a single nest.

A clutch of eggs can be as large as twelve or more if two hens nest together. The large number of eggs and young raised in a season is a survival strategy to counter the high losses in the cold northern winters.

Sufficient and varied live food is essential for breeding this delightful species. Aviculturists who have conservatory-type plant houses have a particular interest in maintaining this very social, small tit species to forage on pest insects and their eggs (aphids).

Color plate 12.8-1
TOP: northeastern subspecies and typical nest.
BOTTOM: central European subspecies.

12.9 Eurasian Tits

azure tit	*Parus cyanus*	**coal tits**	*Parus ater*
blue tit	*Parus caeruleus*	**varied tit**	*Parus varius*

Color plate 12.9-1
TOP LEFT: azure tit.
TOP RIGHT: blue tit.
CENTER: coal tit.
BOTTOM RIGHT: varied tit.

The forty-six species of true tits belong to the family Paridae. They are lively, acrobatic birds, which have very slight or no dimorphism. Male blue tits are more intensely colored.

Tits are found in the woodlands, parks and gardens in Europe, Asia, Africa, India, and North America. Many have adapted to urban environments where they accept birdhouses and winter-feeding stations. They are cavity nesters and nest in vacant woodpecker homes, nest boxes, and other cavities. Outside the breeding season tits travel in mixed species flocks including nuthatches and kinglets.

Tits are very savvy birds, which soon tame down and quickly learn where to find food and who is bringing it to them. I have found coal tits to be most trusting and least afraid of people approaching them.

They are very inquisitive und immediately explore their new surroundings without hesitation. One must take great care to inspect the aviary before release to ensure that it is in good repair. Mesh opening should be no greater than ½-inch (0.12 cm) in diameter. A safety porch is a must and all doors should have self-closing mechanisms to prevent a door staying open for them to escape. Tits must be given sleeping boxes or suitable cavities for the cold season to stay warm at nights. My blue tits prefer suspended, hollowed-out coconuts for sleeping cavities.

The larger azure tit is bred often in European aviaries. It has very attractive shades of blue and white, similar in pattern to the blue tit. The varied tit has been imported from Asia and is sometimes called Japanese tumbler. It too is a delightful species to care for.

Great tits *Parus major* are also bred in aviaries. They are larger and more pugnacious towards other birds. Males have a wider black ventral stripe than hens.

Size: azure tit: 5- inches (13 cm)
blue tit: 4-½ inches (11.5 cm)
coal tit and varied tit are slightly shorter than the blue tit.
Weights range from 9 to 14 gm.

Diet: The same as the diet for the bearded tit and long-tailed tit. I use salt-free margarine or suet mixed with finely shopped nuts and poppy seed as a vehicle to supply vitamins and minerals by mixing in the supplements. Mealworms, buffalos, waxworms and their pupae, and wax moths are readily eaten, however, in my aviary coal tits, blue tits and great tits *Parus major* ignored crickets.

Breeding: Tits have a better chance than open-cup nesters to breed in a mixed aviary with pekin robins or mesias. Blue and coal tits do not inhibit mesias from raise their broods in a common aviary. Tits compete for insects, but mesias select live crickets to raise their young, which tits ignore.

Nest boxes should not have a perch protruding from the entrance hole in order to discourage visits and disturbance by other, less agile birds. Tits do not require a perch for the nest box. The entrance hole should be 1-1/8 inches (29 mm) in diameter. Moss, hair, wool, feathers, and fine plant fibers are used to construct the nest.

Clutch size can be up to twelve and more. Incubation is about twelve days, but the nesting period is longer than two weeks, influenced by available food supply and number of chicks. Two broods per year can be expected in an aviary environment. My blue tits preferred waxworm, buffalo worms and moths (UV light trap) over mealworms to raise their chicks.

Tits are best kept as single pairs for breeding in smaller aviaries.

12.10 Bulbuls

red-eared bulbul	*Pycnonotus jocosus*
red-vented bulbul	*Pycnonotus cafer*
white-cheeked bulbul	*Pycnonotus leucogenys*
black-crested bulbul	*Pycnonotus melanicterus*

Bulbuls belong to the family of Pycnonotidae. Some 120 species are found in the Old World, mainly Africa, Middle East, Japan, and South East Asia. Asian species often found in aviculture are the red-eared or red-whiskered bulbul, red-vented bulbul, white-cheeked and black-crested bulbul, and others.

Most bulbuls are rather shy and docile birds, which live in treetops down to the lower shrub level. The red-eared bulbul is an outgoing species and not as shy as others. They are gregarious outside their nesting season and at times associate with babblers and starlings to form bands of birds.

Some bulbuls have pretty songs; all vocalize a lot. It is a popular cage bird in its homeland as well as in aviculture abroad.

The color of the sexes is alike; juveniles are duller and have less distinctive markings. Two birds may "buddy up" like pekin robins and thus mislead the keeper into thinking they are a pair.

The red-eared bulbul has become accustomed to human habitation, where it feeds on fruit and insects in gardens and plantations.

Bulbuls are hardy birds, easy to acclimate and care for. Winter shelter is recommended for freezing weather.

Size: red-eared bulbul: 8 inches (24.4 cm)
red-vented bulbul: 9 inches (23 cm)
white-cheeked bulbul: 7 inches (18 cm)
black-crested bulbul: 7-½ inches (19 cm)

Diet: The same softbill food as the pekin robin, but more fruit and fewer insects are required. They are fond of flying insects and crickets. Bulbuls eat almost anything. They are fond of fresh figs and nectar drink.

Breeding: A single pair per planted aviary provides the best chance for breeding them. Mates of a given pair are compatible and can be kept together all year. Other pairs or single birds are driven away from nest sites.

Bulbuls build rather bulky, cup-shaped nests by using coarse plant material, rootlets, and twigs. The hen does most of the work. The nest is lined with finer material such as moss hair, grasses, and seed heads of grasses. Raffia and coconut fiber is offered in aviaries. The nests are hidden in vegetation 3 to 10 feet (1 to 3 m) above ground. Nest boxes with half-open fronts may be accepted and nest supports are suggested to secure the nest. Bulbuls are noted for abandoning their nest if they are unduly disturbed.

Clutches have typically two to three eggs, sometimes four or five. Both parents incubate the eggs for twelve to fourteen days and the chicks fledge after thirteen to fifteen days. The chicks are fed live insects for the first few days and fruit plus other "adult" food after that.

Once the young birds become independent they should be transferred out of the breeding aviary because the adults may become aggressive, particularly if they proceed to nest again.

Color plate 12.10-1
TOP RIGHT: white-cheeked bulbul.
TOP LEFT: red-eared bulbul.
CENTER: red-vented bulbul.
BOTTOM: black-crested bulbul.

12.11 White-eyes

Oriental white-eye	*Zosterops palpebrosa*
Japanese white-eye	*Zosterops japonica*
chestnut-flanked white-eye	*Zosterops erythropleura*

Color plate 12.11-1
TOP: Oriental white-eye.
CENTER RIGHT: Japanese white-eye.
BOTTOM LEFT: chestnut-flanked white-eye.

The white-eyes family Zosteropidae has about eighty-five species, which can be found in Africa, Asia southward to Papua, New Guinea, Australia, South Pacific Islands and New Zealand.

They are monomorphic. They have, with few exceptions, white eye rings, are small birds generally grey, yellow-olive, and brown with black and white markings.

This group of birds feeds on nectar with a brush-tipped tongue. They are lively birds, which do not give much enjoyment in a confining cage, where they hop back and forth relentlessly between two perches. In a planted aviary they are a pleasant addition as they move about in a small flock. Their song is low in volume but varied and frequently presented.

The species is gregarious and engages in frequent mutual grooming. This is believed to bring the flock together and reduce hierarchical stress and conflict. They approach in flock formation, intimidating other larger birds at the feeding station and remain there by feeding in unison to gain dominance.

During breeding season they become territorial and will take on even larger birds. Established pairs may kill other members of their own species if left in their territory.

Size: 4-½ inches (11.4 cm) for the noted species.

Diet: Like that of the pekin robin, plus nectar drink and fruit flies.

Breeding: To obtain a pair requires DNA testing or acquiring several birds and hoping that a pair emerges. This becomes evident when two birds initiate nest building and become intolerant to the remaining birds. Because of their gregarious nature they may, however, remain as a group and not nest.

I managed to breed the Oriental white-eye by isolating two birds and watching for courtship behavior. They accepted a small aviary 3 x 5 x 9 feet high (0.9 x 1.5 x 2.74 m) and built their nest three feet above ground in a young hemlock fir. They used dry grasses and some raffia to construct the nest bowl and lined it with finer grasses. The nest seemed hastily built and it failed after the third egg was laid. A nest support was pushed under the nest and the pair had four eggs again nine days later. Three hatched after ten days of incubation. Both parents incubate and brood, and chicks fledge after twelve days incubation. A second clutch was hatched a month later, but abandoned due to disturbance.

The chicks behave like pekin robins after fledging and huddle together within two days roosting body to body on a high branch in the aviary. They became independent after twenty-four days.

The parents were provided with small mealworms (freshly shed when possible), waxworms including their pupae, moths, and small crickets. They showed no interest in whiteworms *Enchytraeus* sp. and did not make much use of an active fruit fly *Drosophila* sp. culture in their aviary, presumably due to the plentiful supply of more "convenient" insects.

Note: Some states of the U.S.A. prohibit the keeping of white-eyes in aviculture.

12.12 Eurasian Thrushes

orange-headed ground thrush	*Zoothera citrina*
white-throated thrush	*Zoothera citrina caynotus*
Siberian thrush	*Zoothera siberica*
European song thrush	*Turdus philomelos*
European blackbird	*Turdus merula*

The orange-headed ground thrush, or dama, is placed with true thrushes, flycatchers, and their allies into the family Musicapidae.

This species can be found in several sub-species ranging from Pakistan, the Himalayas, Southern China, and southward to Thailand. They live singly or in pairs and forage mostly on or near the ground in dense forests, grasslands and cultivated lands. They are dusk and dawn active (crepuscular) and prefer to stay hidden in the cover during the day.

The species is dimorphic. The orange-headed ground thrush is easy to care for and a good singer.

I have kept the gentle and shy orange-headed ground thrush (or dama thrush) with pekin robins, red-tailed minlas *Minla ignotincta*, and European robins *Erithacus rubecula* in a fairly small aviary (7 x 10 x 8 feet high [2 x 3 x 2.5 m]) and found that this relatively large bird — of the size of the American robin *Turdus migratorius* — to be very compatible, however none of the species in the mixed group set up nesting territories during cohabitation.

Size: 8-¼ inches (21 cm)

Diet: Softbill food mix, insects, earthworms, small snails and slugs, plus fruit and berries.

Breeding: Captive breeding has been achieved in planted, spacious aviaries. It is best to keep one pair to an aviary for breeding. The male and female do not need to be separated outside the nesting period as must be done with some other members of this family.

They build a bulky nest much like the American robin or European blackbird and, like them, use mud to plaster the inside of the nest bowl. This must be given consideration by establishing a mud patch in the aviary during nest construction. Thrushes accept open-fronted nest boxes.

The chicks are raised with earthworms, besides other live food. It is helpful to maintain more than one compost pile for the harvest of earthworms, and even to set one up inside the breeding aviary.

The Siberian thrush is another Asian species seen in aviaries. The European blackbird and the European song thrush are cared for and bred in the same way as the orange-headed ground thrush. The European species are rather large, 11 inches (28 cm), which must be taken into account in selecting the size of their aviary. They are however popular in North America with bird keepers who grew up in Europe.

The European blackbird may not be kept in all states in the US, in light of the possibility of its escape and establishment as an alien species feared to damage fruit and berry crops.

European blackbirds and other thrushes are beautiful singers in the wild (*in situ*), but not necessarily in captivity when they were raised without the benefit of hearing the songs of wild mentors.

Color plate 12.12-1
TOP RIGHT: white-throated thrush.
CENTER LEFT: female dama.
CENTER RIGHT: male dama.
BOTTOM LEFT: European song thrush.
BOTTOM RIGHT: European blackbird (*left*) and Siberian thrush (*right*).

12.13 Eurasian Robins, Bluethroats, Redstarts

European robin	*Erithacus rubecula*
white-spotted bluethroat	*Luscinia svecica cyanecula*
red-spotted bluethroat	*Luscinia svecica svecica*
Siberian rubythroat	*Luscinia calliope*
redstart	*Phoenicurus phoenicurus*

Color plate 12.13-1
TOP LEFT: female redstart. TOP RIGHT: male redstart. CENTER LEFT: female and male bluethroat *L. s. cyanecula* and male bluethroat *L. s. svecica*. CENTER RIGHT: female and male Siberian rubythroat. BOTTOM CENTER: European robin at its nest.

The above species are grouped together because of body size and their husbandry. They all belong to the family Musicapidae, sub-family Turdidae. They are smaller than true thrushes and live in Europe and Asia. All are popular birds in aviculture, particularly in European aviaries. The chest spot of the bluethroat varies in subspecies. The Scandinavian/Russian form has a red spot, while the southern and central European population shows a white spot.

Management and breeding is comparable to pekin robins, with the important difference that these species live in solitary, such that they cannot be housed together in one aviary, unless it is enormous in size and habitat complexity, or a pair has formed and is nesting. The European robin is monomorphic and DNA sexing is the most practical way to establish the sex. The only clue may be the wider band of red on the forehead of the male. The other three species are dimorphic. The birds are hardy, but shelter should be given for wet and severe cold weather.

Size: 5½ inch (14 cm). The rubythroat is slightly larger.

Diet: The same as that of the pekin robin; less interested in crickets but accepting earthworms and grubs.

Breeding: Adults must be given a their own aviary unless they are paired up during the breeding season. Aviaries which house single males and females should have connecting sides to monitor the onset of courtship to determine if the birds are ready to accept each other.

With European robins, the hen seeks contact with the male and approaches the mesh wire divider. The male will attack her to drive her from his territory, but if he is in a reproductive mood he will display by raising his head, ruffle his throat feathers, and sing. The hen responds by stretching upwards, dipping excitedly, and showing her ruffled under-tail coverts. The birds can then be tried together and if the male does not chase the hen aggressively, the pairing is underway. The male may also feed the hen during courtship.

The male should be introduced to the hen's aviary, which should be the larger, more complex (planted) breeding habitat, to give the hen the advantage of knowing all escape routes in case the male is still in an aggressive state.

It takes experience to "read" the species-specific courtship behavior. For example, the redstarts display vertical, short courtship flights in which they face each other. The introduced birds must be watched carefully to confirm obvious signs that both are ready to harmonize or to be separated again due to erupting territorial aggression. Once the pair begins to build a nest they can be left together until the breeding season ends and the solitary, territorial behavior returns.

These species are unforgiving to other members of their species in their nesting enclosure. Confrontations will, in all likelihood, end fatally.

Cup-shaped nests are built in closed or half-open nest boxes, crevices, and dense vegetation, depending on species. Clutch size is three to six eggs, and incubation is around twelve days, with a similar nesting period. More than one nesting is expected in *ex situ* management.

Juveniles need to be taken from the breeding aviary as they become independent.

12.14 White-rumped Shama   *Copsychus malabaricus*
Oriental magpie robin or dayal   *Copsychus saularis*

Shamas and dayals occur in several subspecies in India, southern China, and Indonesia and islands. Dayals are also found in Sri Lanka and shamas have been introduced to Hawaii. They live in the shrub level of forests from lowlands to hill country up to 5,000 feet (1,500 m) in elevation.

They are coveted songbirds in their homelands and in aviculture. Both sexes sing. They are solitary birds and highly territorial, which makes it necessary to keep adults in individual enclosures except for the period of nesting and rearing chicks. Breeding solitary species requires more space and management effort.

Size: shama: male – 11 inches (28 cm); female – 8-¾ inches (22 cm)
       dayal: 8 inches (20 cm)

Diet: Softbill food as described above for pekin robins, however finely ground and mixed with hard-boiled egg and insects, especially buffalo worms. Some breeders use moistened chick starter and trout chow, puppy chow, or mynah pellets. The food mix must be moist, but crumbly. Live insects should be a substantial part of their diet. I found that they consume crickets all year around and prefer juvenile over adult crickets.

Breeding: Shamas and dayals can been bred in captivity, provided a single compatible pair is set up in a breeding aviary. While it may be possible to hold other unrelated species in their aviary, more consistent breeding and successful rearing of young is achieved if they are left to themselves.

The introduction of potential mates is done as noted with the European robin (12.12). Since shamas and dayal accept being held in a small cage the male can be brought into the hen's enclosure to test the courtship response and timing to leave them together. This process requires experience and it is better to err on the safe side than risk injury to the hen, or the male if he is the weaker bird.

The courtship is impressive and lively accompanied with much vocalization. A half-open or closed nest box, 6 x 6 x 8 inches high (15 x 15 x 20h cm) with an opening of about two inches (5 cm) or larger is used by the hen to build the nest out of plant fiber rootlets and, in particular, coconut fiber. The nest may be near ground level to about six feet up.

The hen lays four to five, and sometimes three or six, eggs and does all the incubation for eleven days. It can happen that the male hassles the incubating hen and needs to be shifted out of the aviary until the chicks hatch. The male will, however, help feed the chicks. Nesting period is twelve to thirteen days.

Young become independent after about four to five weeks of age. Juveniles must be moved as soon as they become independent and should be separated from each other when they have replaced their juvenile plumage. Breeding pairs are separated after the nesting season.

Color plate 12.14-1
TOP LEFT TO RIGHT: juvenile, male, and female white-rumped shama. BOTTOM: Oriental magpie robin.

12.15 Old World Flycatchers

Asian paradise flycatcher	*Terpsiphone paradisi*
verditer flycatcher	*Musicapa thalassina*
Japanese blue flycatcher	*Cyanoptila cyanomelaena*
rufous-bellied flycatcher	*Niltava sundara*

Color plate 12.15-1
TOP RIGHT: Asian paradise flycatcher.
CENTER LEFT: verditer flycatcher.
CENTER RIGHT: Japanese blue flycatcher.
BOTTOM: rufous-bellied flycatcher.

Many of the Asian flycatchers have lovely blue colors in their plumage. Examples are the Japanese blue flycatcher, the rufous-bellied flycatcher and the verditer flycatcher.

These species are dimorphic. They have flat triangular bills with long bristles at the gape, short legs, and small feet. They are mainly insectivorous. Their voice is pleasant and many incorporate parts of songs of other species in a mocking bird fashion.

Many live in mountain forest at high elevations, which makes them hardy birds in aviculture once they have been acclimatized and accept an artificial diet. Acclimation is often difficult and none of the species are regularly available to aviculturist. Flycatchers are very attractive aviary birds, which get along with other species unless they are nesting.

Aviary breeding has been uncommon so far. Flycatchers are not good prospects for propagation for the novice bird keeper. More attempts should be made to breed these beautiful and interesting birds in our aviaries.

Size: Paradise flycatcher: 8-½ inches, plus 10-inch tail length; more for male (20 cm plus additional 25 cm for the male).
Other species: 7 inches (18 cm). The verditer flycatcher is slightly smaller.

Diet: Like shamas, flycatchers may accept some small berries and chopped raisins or other sweet dried fruit. They consume berries in the wild during the winter season.

Breeding: A well-planted aviary given to a single pair offers best chances to breed them. These birds are solitary but less aggressive towards each other than other species noted above. There may be more confrontations prior to nesting when the birds claim their territory, at which time they may need to be separated and reintroduced again later.

Half-open nest boxes and nest supports, suggesting a nest bowl, should be placed in concealed places to accommodate their secretive nesting behavior. The nest is built out of fine rootlets, dry plant fibers, moss, lichen, and fine grass. Clutch size is three to five eggs, incubation twelve to fourteen days, and the nesting period slightly longer. Independence is reached at five to six weeks of age.

Small soft-bodied insects are particularly important for the first days after hatching. Housefly cultures can be set to produce gut-loaded flies inside the aviary (chapter 9.7). Moths, killed just before feeding, can be harvested from wax moth cultures and trapped with an UV light trap (chapter 9.10). Fruit fly cultures are further options to provide suitable food for the chicks.

12.16 Leafbirds and Fairy Bluebird

greater green leafbird	*Chloropsis sonnerati*
golden-fronted leafbird	*Chloropsis aurifrons*
orange-bellied leafbird	*Chloropsis hardwickii*
Asian fairy bluebird	*Irena puella*

Leafbirds, also called fruitsuckers, belong to the family Irenidae. The Asian fairy bluebird *Irena puella* is another popular aviary bird of this family.

Leafbirds live in the upper layer (canopy) of dense forests in India and Southeast Asia. They occur in pairs and small flocks in search of food and can be seen visiting fruit plantations.

Their plumage color blends well into the green jungle. The golden-fronted leafbird is almost monomorphic but most others are not. The orange-bellied leafbird and the greater green leafbird are dimorphic.

Leafbirds are compatible with other hardy birds not smaller than the pekin robin. They are at times aggressive to their own kin in the aviary environment, including their mate; breeding can be difficult for that reason. There must be particularly good plant cover for the hen to escape the male's aggression. Adjoining aviaries with shift doors are helpful to separate and reintroduce the birds without the need of capturing them.

They are fine singers and like to mimic other birds. Leafbirds become tame and can live for more than ten years.

Size: 7 to 8 inches (18 to 20 cm). The fairy bluebird is 10 inches (25 cm).

Diet: Fruit, softbill mix moistened with mashed fruit and juices, hard-boiled egg, and few insects. Like white-eyes they feed on nectar with their brush-tipped tongue. Nectar drink can be purchased in powder form or mixed by dissolving two tablespoons (15 ml) of fructose in half a cup (100 ml) of warm water. Vitamins can be added to it. It is not important to add protein concentrate as it is done for hummingbirds since the general diet contains sufficient amounts of protein. Halved oranges and grapes can be placed in their aviary.

Breeding: Breeding has only been achieved on rare occasions. The aggressiveness between mates requires watchful management. A well-planted aviary with good ground cover helps to reduce confrontations. Only one pair per aviary should be tried.

The well-hidden nest is constructed relatively high in a fork of a tree or a dense tangle of branches with long strands of plant fiber, rootlets, sisal, raffia, and coconut fiber. Nest supports can be offered.

While males participate in the nesting duties, some will continue to make occasional attacks on the hen. It may be necessary to remove the male from the aviary for a period of time. Two to three eggs are the norm. Incubation period is thirteen to fourteen days and the nesting period is about the same. Two to three clutches may be raised in a season under good aviary conditions.

Both parents feed the young. The chicks are fed with live or freshly killed insects. Freshly shed mealworms or their pupae, buffalo worms, waxworms, moths, and young crickets are offered. In one recent report orange-bellied leafbirds showed a preference for propagated cockroaches. As the chicks become independent they consume fewer insects.

Color plate 12.16-1
TOP LEFT: Asian fairy bluebird.
Top far right: greater green leafbird.
CENTER RIGHT: golden-fronted leafbird.
CENTER LEFT: male orange-bellied leafbird.
BOTTOM LEFT: female orange-bellied leafbird.

12.17 Tanagers, Euphonias, Chlorophonias

blue-winged mountain tanager	*Anisognathus flavinuchus*
superb tanager	*Tangara fastuosa*
Brazilian tanager	*Ramphocelus bresilius*
blue-naped chlorophonia	*Chlorophonia cyanea*
violaceous euphonia	*Euphonia violacea*

Color plate 12.17
TOP LEFT: blue-winged mountain tanager. TOP MIDDLE: superb tanager. TOP RIGHT: Brazilian tanager. BOTTOM LEFT: blue-naped chlorophonia. BOTTOM RIGHT: violaceous euphonia.

Although different in appearance, the above species, along with honeycreepers, belong to the same sub-family, Emberizinae, and tribe Thraupini (tanagers and relatives). Tanagers resemble finches in their bill shape, while honeycreepers look distinctly different. The majority of tanagers are dimorphic.

Buntings, which have an even greater finch-like appearance, are also members of the Emberizinae. Yellowhammers *Emberizia citronella* and other buntings are frequently available to aviculturists, who must recognize their relationship to their soft-food-eating relatives by adding softbill food mix, hard-boiled eggs, and insects to the typical seedeater diet of finches. Live insects are essential for the breeding of this entire sub-family of birds.

Various species of South American tanagers, euphonias and chlorophonias are offered to the aviculturist by importers. Many are colorful and popular birds.

About a third of about 160 South American tanagers live in the Andes and are rather hardy birds of the genera *Thraupis*, *Ramphocelus* and *Anisognatus*. They spend most of their time in the forest canopy searching for fruit, nectar, and insects. These robust species require only shelter in the winter from severe cold, drafts, and particularly wet conditions. Members of the genus *Tangara* live at lower elevations and need more warmth during the cold season with temperatures above 60°F (16°C).

The housing requirements for any species must be researched prior to acquisition. In nature tanagers are gregarious outside the nesting season and form small flocks that may include other species of tanagers. Most are generally compatible with birds of their own size except during nesting.

Euphonias and chlorophonias are from the warmer forest regions; they are smaller than tanagers and more delicate and best kept at room temperature. Sexes are dimorphic with few exceptions.

Size: Most tanagers 6 to 7 inches (15 to 18 cm).

Diet: The same as for leafbirds plus moistened egg cake and some chopped greens or vegetables. Tanagers occasionally eat, and/or damage, foliage of plants in their aviary.

Breeding: A pair should be given its own well-planted aviary of the size noted for pekin robins. The hen builds the nest in dense bush or tree cover. Nest supports and open-nest boxes are accepted. Nest material is like that of other softbills above. Two to three eggs are incubated by the hen for twelve to fourteen days.

The chicks are fed small and soft-bodied insects for the first few days; later, larger mealworms and waxworms can be offered. The live food should be prepared with vitamins and minerals. The chicks fledge after two weeks and are cared for by the parents for two to three more weeks. The hen will nest again before the first clutch is weaned, but the male continues to feed the fledged young. They should be moved from the aviary before the next clutch fledges (chapter 6.13).

Euphonias and chlorophonias are more difficult to breed than tanagers. The birds seen in aviculture are mostly wild-caught specimens and breeding in *ex situ* is an important goal.

12.18 Honeycreepers

red-legged honeycreeper *Cyanerpes cyaneus*
purple honeycreeper *Cyanerpes caeruleus*
green honeycreeper *Chlorophanes spiza*

There are five species of honeycreepers of which the three listed above are commonly found in aviculture. In addition there is the short-billed honeycreeper *C. nitidus* and the shining honeycreeper *C. licidus*, both of which appear to be uncommon in aviculture. The three illustrated species have been bred repeatedly. The author did not find reports of breeding the short-billed or shinning honeycreeper *ex situ* and assumes, therefore, that this is rare.

Honeycreepers are tropical New World birds with long, curved beaks. They are interesting and colorful birds that do well in aviaries. The males and females have different plumage. The leg coloration is different between sexes in the red-legged honeycreeper.

Longevity may be ten years and more. Honeycreepers require warm winter shelter, but can be housed and bred in outdoor aviaries for the summer season. They can be housed with similar sized birds, except for the green honeycreeper, which tends to be more aggressive.

Size: Most are 4 to 5 inches (10 to 13 cm).
The short-billed honeycreeper is smaller at 3-½ inches (9 cm) due to the length of the beak.

Diet: Proprietary nectar drink; soft, sweet fruit; egg cake; finely ground softbill mix, enhanced with small mealworms; waxworms; small crickets; fruit flies; buffalo worms; plus vitamin-mineral supplement.

Breeding: For breeding a pair should be given its own well-planted aviary. The hen builds the nest and uses plant material and cobwebs. Nest supports should be provided since some nests are not very solid. Half-open nest boxes may be accepted as well.

Usually only two eggs are incubated for twelve to thirteen days. Honeycreepers may have two or more broods per breeding season. Ample small-sized, live food must be given to the parents to raise their brood. Nectar and soft sweet fruit is fed to the chicks as well as they mature.

Weaned young must be removed from the breeding aviary as the parents become aggressive towards them, especially if subsequent clutches are hatched.

Color plate 12.18-1
TOP RIGHT: female red-legged honeycreeper.
TOP LEFT: male red-legged honeycreeper.
BOTTOM LEFT: male purple honeycreeper.
BOTTOM MIDDLE: male green honeycreeper.
BOTTOM RIGHT: female green honeycreeper.

Glossary

aberrant behavior: Nontypical, often dysfunctional behavior.

acclimation (acclimatization): Gradual adjustment to new environment including diets.

adaptation: Advantageous changes of an organism to changes in its environment.

air sac mites, *Cytodites nudus*: A parasite living in the air sacs of birds.

air sacs: Thin-walled, bladder-like structures connected to the respiratory system. They aid in internal ventilation to regulate body heat in all birds and buoyancy in aquatic species, but are not involved in gas exchanges (CO_2 and O_2). Typically birds have four pairs of air sacs.

allopreening: Mutual preening of inaccessible body regions by a bonded bird or birds in a bonded flock. The behavior strengthens bonding. *See also* buddy relationship.

altricious/altricial: Young hatched in helpless and immature condition, unable to leave the nest at hatching; also called nidicolous.

amino acids: Organic compound forming an essential part of a protein molecule.

ambient temperature: Temperature of the surroundings.

Annelida: Taxonomic term for segmented worms (earthworms), a phylum of the Invertebrates.

anting: A bird deliberately exposing its plumage and skin to ants for reasons not entirely understood.

apiarist: A beekeeper.

arthropods: Invertebrate animals belonging to the phylum Athropoda including crustaceans, arachnids (spiders, scorpions, mites), and insects.

articular gout: Gout of the joints as opposed to visceral gout, which is that of lining tissue.

as-fed: A term used in analyzing a diet meaning the food mass as presented (to the bird).

aspiration: The accidental breathing in of food particles or drops of fluid (fluid and food aspiration). Often causing aspiration pneumonia.

avian flu: Strains of influenza viruses affecting birds, and which can also be zoonotic.

aviary: A flight cage for birds of various dimensions. In the context of this book a bird enclosure large enough to walk into.

babblers: A group of social birds which usually communicate with a "babbling" contact call; taxonomically belonging to the family Sylviidae, tribe Timaliini.

bacterium flora: A symbiotic and beneficial group of bacteria in the gastro-intestinal (GI) tract, which aid digestion.

banding: The application of leg bands or rings to a bird's leg for identification.

biomass: The total content of organisms in an environment.

biotope aviary: An aviary that is designed to simulate a biotope or habitat type.

biotope: Small region with uniform environmental conditions and organisms.

birdcage: An enclosure to contain a bird. In the context of this book, it is a small cage that can be carried by hand.

bismuth subsalicylate: A human liquid medication for indigestion, which provides a special coating action to give relief in cases of diarrhea in birds.

blood feather: A feather in the process of regrowth while the quill is still soft and supplied with blood and serum.

box cage: A cage with only one side (usually the front) open for viewing and servicing.

breeding consortium: A group of individuals engaged in the collaborative propagation of animals.

buddy relationship: The bonding of two birds, including same sex couples, of the same or mixed species, which become partners in allopreening and roosting.

buffalo worms, *Alphitobius diaperinus*: The lesser mealworm.

bush fly, *Musca vetustissima*: A species of fly found in Australia and widely used for artificial cultivation by Australian aviculturists.

calcinosis: Calcification of soft tissue.

Calgary Zoo, Botanical Garden and Prehistoric Park: A major zoological and botanical garden in Calgary Alberta, Canada (www.calgaryzoo.org).

Canthaxanthin: A carotenoid, which is produced by industrial method as a food coloring agent and diet supplement to enhance red colors in a bird's plumage.

capped/capping: When the hatching chick makes a circular cut to open the eggshell so it can emerge.

carbohydrates: Compounds containing carbon combined with hydrogen and oxygen, i.e. sugars, starches and cellulose.

carotenoid(s): Yellow to red lipochrome pigment (biochrome) found in plants and animals, which enhances yellow, orange, and red plumage coloration.

cave crickets, *Ceuthophilus* **spp.:** Also called camel cricket, a ½-inch (12 mm) wingless cricket.

cecum/ceca (also spelled caecum, caeca): Blind gut(s) which support a bacterium flora to aid the digestion of cellulose in birds and other animals.

cellulose: A long-chain polymer polysaccharide carbohydrate, of beta-glucose. It forms the primary structural component of plants and is the most abundant form of living terrestrial biomass.

chitin: A colorless, hard, amorphous substance (polymer) that forms the exoskeleton of insects and crustaceans.

chlorophyll: Green nitrogenous substance of chloroplast in plants essential for photosynthesis to produce carbohydrates.

CITES (Convention in International Trade in Endangered Species of Wild Fauna and Flora [1975]): An international treaty signed by over 100 nations to protect endangered species trough a permit system for commercial import and export. Species are listed by Appendices I, II, and III, depending on their status in the wild.

cloaca: Common passage region for feces, urine, sperm and eggs leading to the anus (vent).

closed leg bands (or rings): Are of a prescribed inside diameter to be applied to a chick's leg (tarsus) prior to hatching when the ring can still be slipped over the foot joint. They cannot be applied later or removed without damaging the foot, and serves as permanent ID. Rings are issued by the ring registrar of an avicultural society.

clutch: The eggs laid in the nest for incubation.

contact species: Species that roost body-to-body and often engage in allopreening.

crane flies, *Tipula* **spp.:** An insect that resembles a large mosquito (> 1 inch (25 mm)).

critical fight distance: The distance at which an animal attacks if it is cornered.

critical flight distance: The distance at which an animal takes flight or flees from a perceived enemy.

crop: A storage organ, a dilation of the esophagus typical for most seedeaters.

deciduous: Plants species dropping and replacing their leaves, versus evergreens that do not.

Dexamethasone: A drug used for anti-shock treatment to reduce local pain and impact trauma.

dimorphic/dimorphism: Difference of form and or color between males and females within a species.

DNA testing (deoxyribonucleic acid): A method to establish sex by examining the chromosome bands. Feather follicle tissue or blood/serum are submitted for the laboratory test.

dragonflies: A large order of insects (Odonata) ¾-inch to 5 inches (18 – 127 mm) long. There are 5000 species worldwide.

dry matter: The components in a food analysis that remain after it is dehydrated, including organic and inorganic substances.

dyspnea: Labored, open-beak breathing in birds.

earwig, *Forficula auricularia*: Flightless insects about ½-inch (12 mm) long.

egg binding: A condition in which a hen has difficulty expelling her egg.

egg cake: A form of bird cake. *See* chapter 8 for recipe.

egg yolk peritonitis: Caused by the presence of egg yolk material in the body cavity, leading to inflammation and bacterial infection.

electrolytes: A solution, which conducts electricity, used to balance blood chemistry and to hydrate an organism.

enchytraeids, *Enchytraeus* sp.: Whiteworm, a small segmented worm used as food animals for birds and amphibians.

entomologist: One involved in scientific studies of insects.

enzyme: A protein produced by cells initiating and activating metabolism.

epiglottis: The small flap at the base of the tongue that shuts off the larynx (upper part of trachea or windpipe) during swallowing in mammals; birds do not have an epiglottis, but a glottis.

essential amino acids: Amino acids the bird must have to function and grow, and cannot produce itself. They must be provided in the diet: arginine, glycine, histidine, isoleucine, leucine, lysine, methionine, phenylalanine, threonine, thryptophane, and valine.

estrus (oestrous): Sexual cycle culminating in ovulation.

ethology/ethologist: Study of animal behavior patterns; scientist studying animal behavior.

ex situ: Not in its normal wild environment, i.e. a bird in captivity; versus *in situ* in the wild.

exertion myopathy: Muscle burn out causing shock.

exoskeleton: The outer body surface/shell of insects, arachnids, and crustaceans.

extirpated: A species regionally, but not globally, extinct from its historic range.

exto/endo parasites: Parasites witch are found externally or internally, on or in organisms.

fat-soluble vitamins: Integrated and stored with lipids in the body; for example, vitamins A and D.

fecal floatation: Testing of a fecal sample by floating parasite ova contained in it to the surfaces for transfer to a microscope slide (ova identification and count).

fecal sack: The droppings of young chicks in the nest encased by a mucous membrane.

fermentation: Break down of organic compounds by ferments, for example, the conversion of glucose into ethyl alcohol.

ferric chloride: A styptic compound that contracts tissues/blood vessels to stop bleeding.

fledging: The leaving of a nest by altricious chicks.

flight: A term for a bird enclosure larger than a small cage, with one dimension of no less than three feet (1 m).

flight/fight distance: *See* critical fight/flight distance.

founders: Individuals with unique genes in a breeding population.

frugivore: An animal that primarily feeds on fruit.

furniture: Structures, objects, plants, etc. which enhance the quality of an animal enclosure, for example, barriers, sleeping places, nest sites, baths, and exercise structures.

gallinaceous: Birds related to chickens, pheasants, and relatives.

gapeworm, *Syngamus trachea*: A parasite that lives in the trachea causing interference with respiration (gaping).

gaping: Meaning "opening their mouth widely," it is a behavioral response to beg for food, or in courtship imitating juvenile behavior. It also occurs when birds are troubled by something lodged in their trachea, for example in gapeworm infection.

gastrointestinal (GI): Pertaining to the gastrointestinal tract, which includes the stomach and intestines.

gene pool: A genetically managed population of animals for the purpose of breeding and conservation.

genetic diversity: Diversity of genes, parentage, and bloodlines within a species.

geno type: The genetic make-up and characteristics of a specimen or species.

glottis: A slit-like opening in the floor and rear of the bird's mouth. A cartilaginous structure that open and closes the entrance to the larynx / trachea of a bird..

glucose: A monosaccharide carbohydrate found in the form of dextrose in plants and animals.

glycerine: (glycerol) A sweet, oily, colorless alcohol used in medicine, industry and the arts.

gout: Inflammation of the joints (articular gout)

and lining tissues (visceral gout) caused by excess uric acid in the blood (high protein diets) and/or a dysfunctional renal system.

granivore/granivorous: Feeding primarily on grains and seeds.

gut loading: The feeding of certain nutrients, minerals, and vitamins to food animals to improve the nutrition, Ca: P ratio, and vitamin content.

habitat: The environment where a species is normally found (*in situ*).

hand-rearing: The raising of orphaned animals by hand.

hardbill: A bird feeding on hard foods such as seeds.

hardware cloth: Fine woven or welded mesh wire used for screens and aviaries.

hatching container: In the context of this book a small container which hold eggs of food insects and food medium for hatching and the early development stage of larvae and nymphs.

Havahart trap: A live trap manufactured by Woodstream, 69 North Locust Street, Lititz, PA, 175443, USA (www.havahart.com).

hookbill: A term for psittacine birds.

hypervitaminosis: Condition caused by over supply of vitamins.

hypovitaminosis: Condition caused by lack of vitamins.

IATA (International Association of Air Transport): An organization that sets rules and conditions for air transport, for example, bird shipping crate construction.

imago: The adult insect, final, reproductive stage of metamorphosis.

impaction: Blockage of the GI tract by over supply of roughage.

***in situ*:** In its original site or position, i.e. an animal in its natural habitat, versus *ex situ*, not in its normal habitat.

inbreeding: The deliberate or accidental breeding of genetically related individuals with each other.

incubation period: The time span between the start of incubation and hatching/birth.

infra specific: Events occurring within individuals of a species.

infrared lamp: Having a wavelength greater than visible red light; particularly effective for warming surfaces it strikes (heat lamps).

inorganic: Pertaining to compounds lacking carbon but including carbonates and cyanides; the residual content after burning a plant or animal.

insect breeding cabinet: A specially constructed insulated cabinet to maintain constant high temperature to incubate and propagate insects.

insectivore: An animal primarily feeding on insects.

intra-specific: Events occurring between individuals of different species.

ivermectin: An antiparasitic drug that controls skin mites, respiratory mites, and roundworms. It can be used intra-muscular, oral, and topical ("spot on").

iron storage disease (haemosiderosis): A disease caused by accumulation of iron in the cells of liver, kidneys, pancreas, heart muscles, and spleen. Diets should contain <70 pm for sensitive species.

ISIS (International Species Information System): A computerized data system for *ex situ* animals, which monitors current inventories of species in global zoological collections.

keel: The shaft of a feather.

keratin: A compound that forms horn, claws, and nails.

knemidokoptic mite, *Knemidokoptes* sp.: A mite causing crusty tissues (dermatitis) around the cere, eyes, feet and legs.

lacewing, *Chrysopa* spp.: A soft bodied insect species > ½-inch (10 mm) cherished by softbills.

lactase: An enzyme involved in the digestion of milk constituents (lactose).

larvae: The immature form of an animal that is unlike the adult and must undergo metamorphosis.

larynx: Upper part of trachea or windpipe connecting to the mouth cavity.

line breeding: The focused breeding of certain performance or phenotypical features of a population, which generally involves inbreeding (a certain bloodline).

lipids: Greasy, oily organic substances including fats, waxes, and sterols.

lipochromes: Fat based biochromes; *see* carotenoids.

longevity: Length of life.

maggots: Larval stage of flies.

mandibles: Upper and lower jaws or beaks.

mayflies (order Ephemeroptera): An insect with a short lived "sub imago " which sheds skin and wings after only a day and then mates to lay eggs. The larvae live sub aquatic.

meniscus: A crescent shaped body such as the concave cap formed by fluids, for example, by overfilling a vial.

metabolism: Chemical processes taking place in an organism to convert nutrients into body tissues (anabolism) and energy (catabolism).

metamorphosis: A marked change in structure and form of an organism from embryo to adult.

methionine: An essential amino acid.

migratory birds: Birds that seasonally migrate to find suitable food sources.

mimicking: To copy or simulate a process; for example, climate changes in an artificial environment.

myopathy: An ailment caused by the excessive use of muscles.

molt: The seasonal replacement of plumage.

monomorphic: Of the same appearance (phenotype).

nare(s): Nostril(s) in birds.

necropsy: Autopsy (as in human medicine) performed on animals.

nectivores: Animals feeding on nectar.

nidicole (altricial): Hatching, helpless birds that remain in the nest until fledging.

nidifugous (precocial): Birds that leave the nest immediately after hatching; for example, fowl.

night lighting: Setting up low level light sources to aid birds to find a perch if disturbed during the night.

nymph: Larval stage of an insect not involved in complete metamorphosis.

off-exhibit breeding: Term used in zoological gardens for breeding facilities not open to general public visitation.

omnivore: An animal that feeds on a wide variety of foods of plant and animal origin.

oocyst: Thick-walled structure containing sporozoites, the infectious agents passed to other birds by ingestion.

open-beak breathing: Caused by a shortage of breath or a response to bring more air into the respiratory system because of higher oxygen need or a blocked sinus passage.

osteomalacia: Softening and weakening of the supporting bones and bone density caused by insufficient mineralization.

ova: Eggs.

ovaries: Female genital gland that produces ova.

ovipositor: A tubular organ of insects to deposit eggs.

parasite: An organism that lives on the expense of another organism for food and shelter.

Passeriformes: Perching birds.

pathogen(s): A disease producing bacterium or microorganism.

Pedialyte: An electrolyte solution designed for children.

perch-feeder: A small clip on dish to entice a fledged bird to feed on its own.

peristalsis: Contractile muscle movement of a tubular organ to move substances to the point of expulsion.

peritonitis: Inflammation of the peritoneum.

phenotype: The outer (visible) characteristic of an organism.

photoperiod: The duration of day light hours.

physiological(ly): Bodily function of organs of an organism.

phytic acid: A complex of phosphoric acid and sugar that chelates minerals. Ca is of particular concern if food insects have been exclusively been fed on wheat bran.

pinfeather: A feather just beginning to grow through the skin.

pinheads: A term for newly hatched cricket nymphs.

pinkies: A term for fly maggots, but also hairless baby mice.

precocial: Chicks that leave the nest at hatching time covered with down.

probiotics: A prophylactic supplement containing: vitamins, essential amino acids, and beneficial microorganisms to establish a gut flora.

propagation: Breeding.

prophylactic health program: A preventative, proactive management regime to avoid diseases.

pro-vitamins: Substance to promote the formation of vitamins.

psittacines: Members of the parrot families.

pupa/pupation: The metamorphic stage before hatching into the adult insects.

PVC: Polyvinyl acetate compound, a resin used for construction/plumbing materials.

quarantine: Isolation of animals for a set period of time to pass the incubation period of potential or diagnosed contagious diseases.

quick: The life tissue core in a toenail.

raffia: Leaf stalks fiber of a cultivated palm *Raphia pedunculata*.

red roost mite *Dermanyssus* **sp.:** An ecto parasite.

ringing: *See* banding.

ring registrar: A member of an avicultural organization designated to disperse and register leg bands.

safety porch: An unoccupied space (a lock) between the aviary occupied with birds and a general entry from an unenclosed area.

self-feeder: Feeding stations for animals; in the context of this book, containers to hold live insects for birds to catch them as needed.

shift cage: A portable cage that doubles as a shift corridor between two enclosures.

smears: Scrapings of body fluids or exudates for pathology examination.

sodium propionate: A compound that inhibits the development of mold.

softbill: A bird that feeds on soft foods.

SPCA (Society for the Prevention of Cruelty to Animals): An international animal welfare agency.

species specific: Having characteristics that are unique to a species.

sphagnum moss, *Sphagnum* **sp.:** A moss growing in bogs.

stereotypical/stereotypy: Animal behavior term for repetitive, senseless movements usually resulting from stress and anxiety.

stimulus–response pattern: A patterned response to stimuli; for example, the gaping of chicks stimulates parents to feed them.

stress bands: Bands of incomplete growth on feathers caused by nutrient deficiencies and related to stress.

studbook: A registry of animals tracing their lineage.

styptic agent: A chemical that constricts vessels to stop blood flow.

substrate: The surface layer.

symbiotic: A mutually beneficial partnership between dissimilar organisms.

tail coverts: Feathers that cover the start of the tail.

talons: Claws of a bird of prey.

tarsus: The long leg bone (shank) between the foot joint and the tibiotarsal joint.

taxonomic order: The systematic arrangement of animal and plant organisms into kingdom, phylum, class, order, family, genus and species.

tibiotarsal joint: The joint between the tibia and the tarsus.

trace elements: Chemical elements found in very small amount in organisms, which have a significant effect on biochemical processes.

under-represented bloodline: A bloodline or animals whose genes have not contributed to the gene pool of a managed breeding population.

understory: The lower layer of a plant community, i.e. brush layer in a forest.

urate(s): Salt(s) of uric acid, for example, deposits around foot joint of a bird with articular gout.

urea: The white substance in bird droppings; a crystalline compound formed by nitrogenous waste products. Birds do not shed urea by urination as mammals do, but excrete uric acid instead.

UV light: A light fixture that emits ultra violet rays and attracts insects at nights.

UV rays: Ultra violet light, which has a wavelengths beyond the visible violet spectrum, for example, sunrays that convert pro vitamins into absorbable vitamin D.

vent: The region of the cloaca.

vernal equinox: Equinox in the spring of the year.

visceral gout: Gout of the ling tissues of inner organs.

volatile fatty acids: The end products of fermentation in the cecum.

weaning: Separating the young from the care of the parents or parents no longer feeding their maturing young when they become independent.

wild-caught: Animals caught in the wild (*in situ*) for various purposes.

X-ray: Electromagnetic radiation of extremely short wavelength which can penetrate solids and act on photographic plates.

zoonotic diseases: Diseases transmissible between animals and humans.

Bibliography

Abdulali, H. 1993. A Catalogue of the Birds in the Collection of the Bombay Natural History Society. *Journal of the Bombay Natural History Society* 80: 28.

Acorn, J. 1958. *Bugs of British Columbia*. Edmonton: Lone Pine Publishing.

Aeckerlein, W. and D. Steinmetz. 2003. *Vögel richtig fuettern*. Stuttgart: Verlag Eugen Ulmer GmbH.

Baars, W. 1986. *Fruchtfresser und Blütenbesucher*, Band 2. Stuttgart: Verlag Eugen Ulmer GmbH.

Barker, D., M.P. Fitzpatrick, and E.S. Dierenfeld. 1998. Nutrition Composition of whole Invertebrates, *Zoo Biology* 17: 123-34

Bates, H. and R. Busenbark. 1970. *Finches and Soft-Billed Birds*. Neptune City: T.F.H. Publications, Inc.

Bell, E.C. 2001. *Encyclopedia of the World's Zoos*. Chicago: Fitzroy Dearborn Publishers.

Bruun, B. 1970. *The Hamlyn Guide to Birds of Britain and Europe*. London: Hamlyn Publishing Group Ltd.

Cooper, J.E. 2003. *Captive Birds in Health and Disease*. Surrey, Canada: Hancock House Publishers.

Everett, T. H. 1966. *Reader's Digest Complete Book of the Garden*. New York, Montreal: The Reader's Digest Association.

Finke, M.D. 2002. Complete Nutrient Composition of Commercially Raised Invertebrates Used as Food for Insectivores. *Zoo Biology* 21: 269-285.

Finke, M.D. 2003. Gut Loading to Enhance the Nutrient Content of Insects As Food for Reptiles: A Mathematical Approach. *Zoo Biology* 22: 147-162.

Fowler, M.E. 1978. *Zoo and Wild Animal Medicine*. Philadelphia, London, Toronto: W Saunders Company.

Friederich, U. and W. Volland. 1998. *Futtertierzucht: Lebendfutter für Vivarientiere*. Stuttgart: Verlag Eugen Ulmer GmbH.

Grewal, B., B. Harvey, and O. Pfister. 2002. *A Photographic Guide to the Birds of India*. Princeton and Oxford: Princeton University Press.

Guenther, E. 2004. Erlebnisse mit dem Orangebauchblattvogel. *Gefiederte Welt* 128: 364-368

Hediger, H. 1964. *Wild Animals in Captivity*. New York: Dover Publication Inc.

Harrison, C. and A. Greensmith. 1993. *Birds of the World*. New York: Dorling Kindersley Inc.

Heinzel, H., R. Fitter, and J. Parslow. 1998. *Birds of Britain and Europe with North Africa and Middle East*. London: Harper Collins Publishers.

Holland, G. 2006. *Encyclopedia of Aviculture*. Surrey, Canada: Hancock House Publishing Ltd.

Karsten, P. 2000. Pekin Robins: Care and Breeding, Part I. *The Avicultural Journal* 23, no. 1: 33-39. Avicultural Advancement Council of Canada.

Karsten, P. 2000. Pekin Robins: Care and Breeding, Part II. *The Avicultural Journal* 23 no. 2: 21-27. Avicultural Advancement Council of Canada.

Karsten, P. 2001. Puppet Rearing Pekin Robins (*Leiothrix lutea*). *The AFA Watchbird* 28, no. 2: 38-41. Phoenix.

Karsten, P. 2001. Breeding the Oriental White-Eye (*Zosterops palbebrosus*) with the help of nest site supports. *The AFA Watchbird* 28, no. 4: 58-60.

Karsten, P. 2002. Pekin Robins: Information on Their Care and Breeding. *The AFA Watchbird* 29, no. 2: 56 – 68. Phoenix.

Karsten, P. 2002. Raising Waxworms as Food Insects for Birds and Reptiles. *The AFA Watchbird* 29, no. 4: 37-38

Kierschke, S. 2004. *Zuchterfahrungen mit Insektenfressern*. Wilhelmshaven: S. Kierschke, self-published.

King, B. et al. 1993. *Collins Field Guide, Birds of Southeast Asia*. London: Harper Collins Publishers.

Milne, M. and L. Milne. 1980. *The Audubon Society Field Guide to North American Insects and Spiders*. New York: Alfred A. Knopf.

Olney, P. 2005. *The World Zoo Conservation Strategy*. WAZA Executive Office. Bern.

Rosskopf, W. and R. Woerpel. 1996. *Diseases of Cage and Aviary Birds*, Baltimore: Williams and Wilkins.

Sibley, C.B. and B.L. Monroe. 1990, 1993. *Distribution and Taxonomy of Birds of the World*. New Haven, USA: Yale University Press.

Sloss, M. et al. 1978. *Veterinary Clinical Parasitology*, 5th ed. Iowa State University Press, Ames, Iowa.

Steinigeweg, W. 1989. *Sonnenvögel-Chinesische Nachtigall: Pflege, Ernährung, Krankheiten, Vehalten*. München: Gräfe und Unzer GmbH.

Steinigeweg, W. 1998. *The New Softbill Handbook* (English translation). New York: Baron's Educational Series, Inc.

Terres, K. 1987. *Audubon Society Encyclopedia of North American Birds*. New York: Alfred A Knopf.

Vince, M. 1996. *Softbills, Care Breeding and Conservation*. Surrey, Canada: Hancock House Publishers.

Wedel, A. 2004. *Ziervögel*. Stuttgart: Parey Verlag.

Wendt, T. 2002. *Einheimische Singvögel halten und züchten*. Stuttgart Verlag Eugen Ulmer GmbH.

Species Index

- Page numbers in **bold** indicate extended discussion on the subject.
- Page numbers in *italics* indicate illustration/photo of species.
- "bc" indicates illustration on book cover.
- Consult general index for additional listings.

Birds by Family

Aegithalidae long-tailed tits 197, *210*, **211**
Fringillidae finches, tanagers, honeycreepers etc. 25, 197, **228**, *229*, *230*, **231**, bc
Irenidae leafbirds 198, *226*, **227**
Musicapidae thrushes, flycatchers etc. 198, *218*, **219**, *220*, *221*, *222*, **223**, **224**, *225*
Paridae true tits 197, *212*, *213*
Pygnonotidae bulbuls 198, *214*, **215**
Sylviidae, includes tribe: Timaliini babblers 197, **200**, *201*, *202*, **203**, **204**, *205*, *206*, **207**, **208**, *209*
Zosteropidae white-eyes 96–98, 163, 165, 198, **216**, *217*, 227

Birds by Genera/Species

Acanthis cannabina linnet 25
Acanthis flammea cabaret redpoll 25

Accipiter cooperii Cooper's hawk 74, *74*
Accipiter striatus sharp-shinned hawk 74, *74*

Aegithalos caudatus caudatus northeastern long-tailed tit *210*, **211**
Aegithalos caudatus europeaus central European long-tailed tit *210*, **211**

Amisognathus flavinucha blue-winged mountain tanager **228**, *229*

Carduelis carduelis (European) goldfinch 25, 169
Carduelis chloris greenfinch 25
Carduelis spinus (European) siskin 25

Chamaea fasciata wrentit 28

Chlorophanes spiza green honeycreeper *230*, **231**

Chlorophonia cyanea blue-naped chlorophonia **228**, *229*
Chlorophonia flavirostris yellow-collared chlorophonia bc

Chloropsis aurifrons golden-fronted leafbird **226**, *227*
Chloropsis hardwickii orange-bellied leafbird **226**, *227*
Chloropsis sonnerati greater green leafbird **226**, *227*

Copsychus malabaricus white-rumped shama 24, 30–31, 92, 95, 161, 198, *222*, **223**
Copsychus saularis Oriental magpie robin or dayal *222*, **223**

Coturnix chinensis blue-breasted or Chinese painted quail 30, **92**

Cyanerpes caeruleus purple honeycreeper *230*, **231**
Cyanerpes cyaneus red-legged honeycreeper *230*, **231**

Cyanoptila cyanomelaena Japanese blue flycatcher **224**, *225*

Erithacus rubecula European robin 24, 27, 95, 161, 165, **181**, 198, **220**, *221*

Euphonia violaceus violaceus euphonia **228**, *229*

Fringilla coelebs chaffinch 25, 30, 98, 221

Garrulax canorus spectacled jay thrush or hwamei **206**, *207*
Garrulax galbanus yellow-throated laughing thrush **206**, *207*
Garrulax leucolophus white-crested jay thrush **206**, *207*
Garrulax pectoralis greater necklaced laughing thrush **206**, *207*

Irena puella Asian fairy bluebird *226*, **227**

Leiothrix argentauris silver-eared mesia *19, 21, 22, 23, 24, 51, 58, 64, 93, 104, 111, 114–115, 161, 197,* **200,** *201*
Leiothrix argentauris laurinus red silver-eared mesia **200,** *201*
Leiothrix lutea pekin robin or red-billed leiothrix (also called Chinese or Japanese nightingale and hill tit) *17, 18,* **27–29**
Leiothrix l. calipyga Indian red-billed leiothrix **16,** *17,* 28
Leiothrix l. lutea red-billed leiothrix **16,** *17,* 28
Leiothrix l. kwangtungensis Kwangtung red-billed leiothrix **16,** *17,* 28
Leiothrix l.utea yunnanensis Yunnan red-billed leiothrix **16,** *17,* 28

Luscinia calliope Siberian rubythroat **220,** *221*
Luscinia svecica cyanecula white-spotted bluethroat **220,** *221*
Luscinia svecica svecica red-spotted bluethroat **220,** *221*

Malurus sp. splendid wren 99, *99*

Minla cyanouroptera blue-winged minla (siva) *24, 75, 82, 98, 202,* **203**
Minla ignotincta red-tailed minla *24, 75, 202,* **203**
Minla strigula chestnut-tailed minla *202,* **203**

Musicapa thalassina verditer flycatcher **224,** *225*

Niltava sundara rufous-bellied niltava **224,** *225*

Panurus biarmicus Eurasian bearded tit or reedling 28, **208,** *209*

Parus spp. Tits *68, 143, 161, 165, 197,* **212,** *213*
Parus ater coal tit *52, 95,* **212,** *213*
Parus cyanus azure tit **212,** *213*
Parus caeruleus blue tit *98,* **212,** *213*
Parus varius varied tit *121, 213*

Phoenicurus phoenicurus redstart **220,** *221*

Pycnonotus leucogenus white-cheeked bulbul *214,* **215**
Pycnonotus jososus red-eared bulbul *214,* **215**
Pycnonotus cafer red-vented bulbul *214,* **215**
Pycnonotus melanicterus black-crested bulbul *214,* **215**

Pyrrhula pyrrhula bullfinch 25

Ramphocelus bresilius Brasilian tanager **228,** *229*

Serinus canarius canary 13

Tangara fastuosa superb tanager **228,** *229*
Tangara chilensis paradise tanager *bc*

Terpsiphone paradise Asian paradise flycatcher **224,** *225*

Turdus merula European black bird *218,* **129**
Turdus migratorius American robin 109
Turdus philomelos song thrush *218,* **219**

Upupa epos hoopoe 160, 219

Yuhina nigrimenta black-chinned yuhina **204,** *205*
Yuhina castaniceps striated yuhina **204,** *205*
Yuhina flavicolis whiskered yuhina **204,** *205*

Zoothera citrina orange-headed ground thrush or dama *218,* **119**
Zoothera citrina cyanotus white-throated thrush *218,* **219**
Zoothera sibirica Siberian thrush *218,* **219**

Zosterops sp. white-eyes **216,** *217*
Zosterops erythropleura chestnut-flanked white-eye **216,** *217*
Zosterops japonica Japanese white-eye **216,** *217*
Zosterops palpebrosa Oriental white-eye *97,* **216,** *217*

Birds by Common Name

babblers, members of tribe Timaliini 28, 31, 75, 183, 197, **200–208,** *201, 202, 205, 206, 209*
black bird, European *218,* **219**
bluebird, Asian fairy *226,* **227**
bluethroat
    red-spotted **220,** *221*
    white-spotted **220,** *221*
bulbul
    black-crested *214,* **215**
    red-eared *214,* **215**
    red-vented *214,* **215**
    white-cheeked *214,* **215**

canary 13
Chlorophonia, blue-naped **228,** *229*

dama or orange-headed ground thrush *218,* **119**
dayal or Oriental magpie robin *222,* **223**

euphonia, violaceus **228,** *229*

European finches 25
    bullfinch 25
    chaffinch 25, 98, 221
    goldfinch 25, 196
    greenfinch 25
    linnet 25
    redpoll 25
    siskin 25, 98
finches, general 25, 29–30, 40, 63–64, 68, 85, 98, 126, 138, 162, 198, 228
flycatcher
    Asian, paradise **224**, *225*
    Japanese, blue **224**, *225*
    rufous-bellied **224**, *225*
    verditer **224**, *225*

hawk 51, 75, 114
    Cooper's 75, *75*
    sharp-shinned 75, *75*
honeycreeper
    green *230*, **231**
    purple *230*, **231**
    red-legged *230*, **231**
hoopoe 160, 219

laughing thrushes
    greater necklaced *206*, **207**
    white-crested *206*, **207**
    yellow-throated *206*, **207**
leafbird
    golden-fronted *226*, **227**
    greater green *226*, **227**
    orange-bellied *226*, **227**

mesia *19, 21, 22, 23*, 51, 58, 64, 93, 104, 111, 114–115, 161, 197, **200**
    red silver-eared **200**, *201*
    silver-eared **200**, *201*
minla
    blue-winged (or blue-winged siva) *24*, 75, 82, 98, *202*, **203**
    chestnut-tailed *202*, **203**
    red-tailed 75, *202*, **203**

niltava, rufous-bellied (see flycatcher) **224**, *225*

quail, blue-breasted or Chinese painted 30, **97**

redstart **220**, *221*
reedling (see bearded tit)
robin
    American 109
    European **220**, *221*
    pekin 16–*17, 18*, **27–29**
rubythroat, Siberian **220**, *221*

shama, white-rumped *24*, 30–31, 92, 95, 161, 198, *222*, **223**
siva, blue-winged (see minla)
splendid wrens 99

tanager
    blue-winged mountain **228**, *229*
    Brasilian **228**, *229*
    superb **228**, *229*
    paradise *bc*
thrush
    European song *218*, **219**
    greater necklaced laughing *206*, **207**
    Siberian *218*, **219**
    spectacled jay thrush (hwamei) *206*, **207**
    white-crested laughing *206*, **207**
    white-throated *218*, **219**
    yellow-throated laughing *206*, **207**
tit 68, 143, 161, 165, 197, **212**, *213*
    bearded, Eurasian (or reedling) 28, **208**, *209*
    azure **212**, *213*
    blue 98, **212**, *213*
    coal 52, 98, **212**, *213*
    long-tailed tit *210*, **211**
        central European *210*, **211**
        northeastern *210*, **211**
    varied **212**, *213*

white-eye 96–98, 163, 165, 198, **216**, *217*
    chestnut-flanked **216**, *217*
    Japanese **216**, *217*
    Oriental **216**, *217*
wrentit 28, *28*

yuhina
    black-chinned **204**, *205*
    striated **204**, *205*
    whiskered **204**, *205*

Families, birds

Aegithalidae (long-tailed tits) 211
Fringillidae (finches, honeycreepers, tanagers) 25, 28, 231
Irenidae (leafbirds) 227
Musicapidae (flycatchers and related birds) 219–224
Paridae (true tits) 212
Pycnonotidae (bulbuls) 215
Timaliini (babblers) 200–208
Zosteropidae (white-eyes) 216

Softbill species listed in pen/ink drawing page 194, chapter 12

Bombycilla garrulous Bohemian waxwing
Buceros bicornis giant hornbill
Calyptomena viridis green broadbill
Colibri coruscans sparkling violet ear (hummingbird)
Coracias caudate lilac-breasted roller
Cosmopsarus regius golden-breasted starling
Cyanocorax yncas green jay
Ducula aenea Sulawesi green imperial pigeon
Gracula religiosa hill mynah
Hylcyon leucocephala grey-headed kingfisher
Lanius collurio red-backed shrike
Malurus leucopterus blue and white wren
Merops apiaster European bee-eater
Nectarinia violacea orange-breasted sunbird
Pitta guajana banded pitta
Psilopogon pyrolophus fire-tuffted barbet
Ramphastos tucanus red-billed toucan
Sitta frontalis velvet-fronted nuthatch
Sylvia atricapilla blackcap
Tauraco erythrolophus red-crested touraco
Urocolius macrourus blue-naped mouse bird

Invertebrates

Achaearanea sp. house spider 105, 107, *144*–*145*, 166
Acheta domestica house cricket 22, 23, *144*, **148–153**
Achroea grisella lesser wax moth (lesser waxworm) 154
Alphitobius diaperinus lesser mealworm (buffalo worm) 23, **161**
Aplopus sp. stick insect 146
Aranaeus sp. orb weaver *144*, 145
Arion spp. slugs 70, 105, 186

Bombyx mori silkworm 147

Calliphora spp. and *Musca* sp. fly maggots **162–164**, *163*
Camponotus sp. carpenter ant **31**, *144*, 145
Carausius morosus Indian stick insect *144*, 145, 146
Ceuthophilus sp. cave cricket **107**, *107*
Chilecomadia moorei Chilean moth (moth of butterworm) 147
Chrysopa sp. Lacewing *144*, 145

Drosophila spp. fruit flies *144*, **165**
Drosophila funebris and *D. melanogaster* fruit fly spp. 165

Drosophila hyde flightless fruit fly 165

Eisenia sp. Earthworm **165–166**, *165*
Enchytraeus albidus whiteworm (potworm or enchytraeids) 165
Enallagma sp. Bluet *144*, 145

Forficula auricularia earwig *144*, 145

Galleria mellonella greater waxmoth 23, *154*, **154–158**
Gryllus bimaculatus Mediterranean cricket 148
Gryllus pennsylvanicus field cricket *144*, 167

Libellula sp. Skimmer *144*, 145
Locusta migratoria Egyptian locust 146
Lubricus sp. earthworm (see *Eisenia* sp.)
Lucilia sp. or *Phaenicia* sp. green bottle flies 163, *163*

Musca domestica house fly (maggot) 162–163, *163*
Musca vetusfissima Australian bush fly 162

Oniscus sp. sow bug 105

Schistocerca gregaria desert locust 146

Tenebrio molitor mealworm (darkling beetle) *158*, **158–160**

Vespula spp. wasp 70, 105

Zootermophis angusticollis dampwood termite 31, 105, *144*, 145, 167
Zophobas morio "superworm" **160**, *160*

Families, invertebrates

Acrididae (grasshoppers) 145
Heptageniidae (mayflies) 145
Lycosidae (wolf spiders) 145
Noctuidae (owlet moths) 145

Other animals

Hyla spp. tree frogs 72, 163, 167
Hyla arborea European tree frog 163
Mus musculus common house mouse 20, 71
Mustela vison mink 73, *73*
Peromyscus maniculatus deer mouse 71, *71*

Procyon lotor raccoon 74
Rattus norwegicus Norway rat 70, *71*
Rattus rattus black rat 71
Sorex sp. Shrew 73, *73*
Thamnophis sp. garter snake 72, *72*

Parasites

Acaridia sp. roundworm 186, *187*
Capillaria sp. threadworm 186–187, *187, 190*
Coccidium sp. protozoan *187*, **187–190**, 190
Dermanyssus spp. grey and red roost mites *174*, **187**, *187*
Knemidokoptes sp. scaly leg mite 186–187, *186*
Mellophaga chewing louse 186, *186*
Sternostoma tracheacolum air sac mite *174*, 186–187, *187*
Syngamus trachea gapeworm *174*, 187

Plants

Azalea sp. 64
Gualtheria shallon salal 143, *143*
Narcissus sp. 64
Phylostachys bambusoides giant timber bamboo 64
Pseudotsuga menziesii Douglas fir 65
Rhododendron sp. 64
Stellaria media chickweed 132, 143
Taraxacum officinale dandelion 132, 143, 150
Thuja plicata western red cedar 65, *65*
Trevoa trinervis trevo bush 147
Tsuga heterophylia hemlock fir 51, 261
Vaccinium pavifolium red huckleberry 143, *143*

General Index

- Page numbers in **bold** indicate extended discussion on the subject.
- Page numbers in *italics* indicate illustration/photo of subject.
- Definitions of terms and topics are provided in the Glossary.
- Consult species index for additional listings.

AACC (Avicultural Advancement Council of Canada) 111
aberrant behavior (see behavior)
acclimation 29, 41, 63, **87**, 88, 125, 175
acquisition 111, 172, **190**, 196
adaptation 29
Avicultural Advancement Council of Canada 111
air sac mites 186, **187**
allopreening (see also mutual preening) 32, 91, **96**, 122
altricial/altricious (nidicolous) 30
amino acids 130
Annelida (see segmented worms)
Arthropods (see spiders)
aviary *19*, **39**, 24
    biotope *18*, **47**, 208
    breeding *19*, 39, **47**
    layout 47, 49, **55–57**
    release to 34, 40, 86, **89**, 99, 123, 190
Avicultural Advancement Council of Canada (see AACC)

babblers *22, 24*, **28**, 30–31, 75, 183, 197, *201, 202, 205, 206, 209*
bacterial infection 117
banding (ringing) *21*, 76, **111–112**, 113–114, 125, 200
bathing *20*, 31
    birdbath 59
    sunbathing 32
beak/bill *18*, 27, **29**, 46, 84, 107, 121, 177–178, 182, 186
bee larvae 126–127
behavior 14, **29–31**, 33, 44, 76, 82–83, 89, 96
    aberrant 81, **92**, 192
    breeding/nesting *20, 22*, 61, 62, 99, **101–107**
    natural 29, 41, 124, 170
    observation 11, 113
    pair/social *18, 19, 20, 21, 22*, 33, 76, **91**–92, 95
    stereotypical 31, 93
    territorial *20*, 48, 92, **97, 99**, 176
biotope aviary *18*, 47, 208
birds of prey 59–60, **74**–75

bismuth subsalicylate 179
bleeding 34, 182, **185**
blood feather 34
body condition 37, **83**–84
bonding/bonded pair *18, 19*, 76, 92
breathing disorder **178**, 179, 187, 192
breeding **91–114**
    aviary *19*, 39, **47**
    events calendar 113
    loans 172
    statistics 114
bones
    brittle bones (osteomalacia) 184
    deformed leg bones (rickets) *23*, 141, 157, 184–185
brooder construction **120**, 127
buddy relationship *18,19*, 33, 89, 92, 215
buffalo worms (see also lesser mealworms) *23*, **161**, 141, 147, 211, 112, 223, 227, 231
bumble foot 180
bush fly 162

cages
    barred 39–40
    box 41–42
    hospital **43–44**, 109, 122, 175–176 184, 191
    portable holding 42
    shift box 42, **54**, 56, 70
    shipping box 43, **45–46**
calcinosis 132
calcium 104, **107**, 125, 127, 133–135, 141–142, 150, 184–185
calcium-phosphorus (Ca:P) ratio **107**, 147, 185
Canthaxanthin 142
Capillaria worms (threadworms) 187, 190
capturing 67
carbohydrates 130–131
cats 50, 53, 60, **73**–74
cecae 131
cellulose 131

chicks
    banding *21,* 79, **111–112**
    early loss 93, 96, **106**–107
    feeding *20, 21*, **104**–**106**, 110, 123
    fledging *21, 22*, 65, 122, **108**
    hand feeding/rearing *22,* 43, 117–118, **120**
    weaning *22,* 94, 102, **110**
    weight monitoring *22,* 123
CITES (Convention on International Trade of Endangered Species) 11, 29, **80**, 87, 91, 200
cleaning 42–44, 50, 58, 137, 152–153
climate 11, **28**, 61–64, 80, 142
Coccidia 187, 189
condition, judging 81
conservation breeding 40, 117, **169**, 172–173
Convention on International Trade of Endangered Species (see CITES)
coughing 178
crickets *22, 23*, 148–153
    cave 107, 144, 167
    cleaning and handling 152
    culture, starting a *23,* 150
    egg-laying and incubation *23,* 150
    field 148, 167
    house 148
    Mediterranean 148
    propagation, large scale *22,* 153
    rearing containers 148
critical flight distance 48

depression 192
Dexamethasone 183
diarrhea 85, 117, 12, 178–179, 187
Dierenfeld PhD, Ellen S. 8, 139, 140, 239
diet 80, 88, 107, 110, **129–131**, 133–135, 138–140, 181
    dry mix formula 135, **138–140**
dimorphism 27
direct smear technique 189
discharges 44, 84, 88, 178–179
dishes
    bathing *20,* 31, 59
    food 40, 42, **58**, 135, 136
    water 46
DNA testing 27, **34**, 81, 172, 185
drinkers/drinking tubes 59, 88
drinking, increased 187
dyspnea (see also breathing disorder) 178, 192

earthworms 70, 105, 146, **165**, 186
egg *21*
    binding *23,* 99, **183**–184

    cake 62, 110, 140, 123, 130, 135–136, 138–**140**, 142
    hatching, bird 30, 93–94, 102–**103**, 113–114, 123, 126
    laying, bird *21,* 93, **102**
    laying, insect *23,* 150, 163
enchytraeids (see whiteworms)
enclosures (see flights and housing)
epiglottis 117
escape 212
essential amino acids 130
ethology 14, **76**
exertion myopathy (muscle burnout) *23,* 109, **191–192**
ex situ 14, 31, 36, 62, 80, 169–170, **173**
export 27, 29, 33, 36, 40, **80**, 91, 111, 196
eyes 84–85
    disorders 179
    opening of 106

fats (see lipids)
feathers 15, 37, 59
    blood 34
    damaged 27, 37, 42, 45, 59, **82**, 88, 183
    pigmentation *18,* 37, **142**
    ruffled/fluffed 31, 32–33, 83, 175, 178, 183, 187
    shedding *18,* 27, 37, 67
    soiled 82, 84, 179
    stress bands 185
fecal
    flotation technique 179, **190**
    sac 106, 122
    samples 44, **177–178**
feces 63, 111
feeding *23,* 46, **129–143**
    behavior 83–84, 98, 104, **106–107**, 109, 122, 143
    chicks 102, **104**, 106–109, 120
    from self-feeders 58, **105**
    insects **150, 156, 159–161**, 163
    routine 58, 63, 72, **137**
    stations *20,* 31, 54, **58**, 61, 89, 159
    utensils *22,* **136**, 123
feet (see scaly legs and swollen feet)
fight distance 48
Finke PhD, Mark D. 8, 144, 239
fledging *20, 22,* 106, **108**, 122, 191
flight distance **48**, 123
flock management **75**–76, 84, 88, 96, 113, 129, 170, 176
fluid aspiration 117, **121**, 126, 178
food (also see feeding)
    color 142
    cultivating, live *22, 23,* 145–166
    insects, value of 141
    presentation 135

flights (enclosures) 56, 47
fly cultures 162
fractures 182
fruit 29, 46, 130, 135, 137, **143**, 150, 161
frugivores **29**, 30, 129, 135

gallinaceous 30
gaping 106, 120, 121
gapeworm 11, 174, **186–187**
garter snake *20*, 72
gastrointestinal disorders 131, 141, 179
giant mealworms 159
glottis 117–178
gout 132, 141, 162, **180–181**
gut-loading **141**, 157

habitat *18, 19*, 28, **41**, 47–48, 65, 91
hand-rearing *22*, **117–126**
    basic equipment 118
    behavior considerations 124
    protocol 126
hatching
    birds *20*, 30, 93, 96, **102–103**, 113, 114, 123–124, 126, 169, 172
    insects *23*, 149–152, 155–156, 159–161, 164–165
    weight, pekin robin *22*, 123
Havahart trap 72
Hawaii Endangered Bird Conservation Program 8, 126
hawk 51, 75
    Cooper's 74
    sharp-shinned 74
head trauma 183
health care 175–191
    common disorders 177–186
house fly 162
housing (enclosures) *19*, 39–64
    aviary 39, 55–57
    breeding aviary 47
    cages 41–45
    small support enclosure 41
    winter 61
hyper aggression 192
hyperactivity 192
hypervitaminosis 132

IATA (International Air Transport Association) 45
ID 171
importation 29, 79–**80**, 87, 91, 112
immune system 80, 104, 117, 132, 175, 178, 186, 192
incubation **103**, 94, 102–104, 114
indigestion 192

infection 85, 86, 107, 132, 174–175, 178–179, 186, 189
    bacterial 117
    digestive system 82, 121
    fungal 84, 187
    respiratory system 84, 180
inflammation 184
insectivores **29–30**, 76, 92, 138, 221
insects 145
    breeding cabinet *22*, 147
    cricket cultivation *23*, 145
    larvae *23*, 154–164
    wild, collecting 166
    trap, UV *23*, 167
International Air Transport Association (see IATA)
International Species Information System (see ISIS)
International Zoo Yearbook (London Zoological Society) 171
intestinal worms 187
iron storage disease 135
ISIS (International Species Information System) 170–172
IUCN (World Conservation Union) 171
Ivermectin 177, 181, 187, **188**

Karsten, Karen 8, 188
Kuehler, Cyndi 8, 126, 127

lameness 141, **180–181**
larvae 145–146
    buffalo worms *23*, 161–162
    butterworms 147
    fly 162–164
    mealworms *23*, 158–160
    parasite 189
    silkworms 147
    superworms *23*, 160
    waxworms *23*, 154–157
larynx (windpipe) 117, **178**
legs 15, 82, 180, 186
    rickets *23*, 107, 141, **184–185**
    rough scales on 180
    thickened/swollen 180, 181
lesser mealworms (see also buffalo worms) *23*, 161–162
Lieberman, Alan 8, 126–127
lifespan (longevity) 31, 36, 40, 117, 140, 231
lighting 62, 75
lipids (fats) 131–132
live food, cultivation of *22–23*, 145–166
longevity (see lifespan)
London Zoological Society (International Zoo Yearbook) 171

loss of
    appetite 107, 179
    birds 106, 193
    condition 109, 175
    weight 123

maggots 105, 146, 163–164
mandible 15, 29, 181–182
    misshapen 184
mating 86, **102**
mealworms *23*
    culture, starting 159–161
    mite infestation 160
mesh (wire) 42, 49, 50, 53, 59, 69, 71–73
metabolism 37, 107, **130**, 132, 138
mice 49, 58, **71**, 20
minerals 37, 96, 121, 130, **133–135**, 138
mink 73
mites 190
    air sac 186–187
    infestation 148, 152, 160
    roost mites 187–188
    scaly leg 187
mollusc (see slugs)
molt *18*, 27, 37, 93, 134
    fright 68, 82
monomorphic 82, 204
mouse *20*, 72
    common house 71
    deer 71
moving (locomotion) abnormally 31, 93, 109, 183, 192
    patterns 60, 64
muscle burn out (see exertion myopathy)
mutual preening (see also allopreening) 32, 89, 91, 96

nares 15, 84, 179
nectivores 29
nesting *20*, 47, 61–62, 65, 91, 99
    material 100–102
    season/period 93, 103, 114
    supports *20*, 101–102
    territory 195
net 32, 43, 51, 54, **67–69**
nidicolous (see altricial)
nidifugous (see precocial)
night lighting 75

obesity 49, 50, 53, 131, 140
observation 51, **76–77**, 88, 98, 113, 177, 193, 196
open-beak breathing (see breathing/dyspnea)
osteomalacia (see brittle bones)
owls 60, 75

paint 44, 46, 49, 50, 53
parasites 70, 84, 175, 179, **186**, 188
    air sac mites 187
    Capillaria (threadworms) 186–187, 190
    Coccidia 187, 190
    detecting presence of 179, **189–190**
    external (ecto) 82, 186, 190
    gapeworm 187
    internal (endo) 186–187
    intestinal worms 186–187
    roost mites 187
    scaly leg mites 187
paresis 183
Passeriformes 28, 196, 199
Pedialyte 126–127
perches 42, 44–45, 47, 52, **59**–60, 65, 75, 108–109, 180, 191
perch-feeder *22*, 123
phenotype **16**, 33, 169
photo period 93, 192
plants *19*, 49, 51, **63–65**
plexiglass 44
plumage **15**, *18*, 27, 31, 37, 82
poisonous plants 63
potworms (see whiteworms)
precocial (nidifugous) 30
predators **70**, 50, 53, 59, 75, 182, 192
    birds of prey 74
    cats 73
    garter snake *20*, 72
    mink 73
    rodents 70–72
    raccoons 73
    shrew 73
    wasps 70, 105
preening 32–33, 91, 96, 122. 125
    over preening 93, 186
probiotics 179
protein 30, 37, 70, 96, **130**, 138–141, 161–163, 180–181
pro-vitamins 132–133, 185
Psittacines 29, 41, 64
psychological needs 33
puppy chow 139, 150, 161

quarantine 41, 79, 80, **84–87**, 112, 186
quick (of toenail) 34

raccoons 73
rat 71
    black 71
    Norway 71

recipe
    dry mix 139
    egg cake 140
record keeping 84, **113**, 169, 171
release 34, 40, 86, **89**, 95, 123, 176, 190
reproduction 11–12, 138, 170
respiratory disorders 177–179
rickets and related disorders *23*, 107, 141, **184–185**
ringing (see banding)
rodents 50, **70–71**, 186, 192
roost mites (see mites)
rubber beak (see rickets)
ruffled feathers (see feathers)

safety porch 47, **54**, 56–57, 212
San Diego Zoo
    Hawaii Endangered Bird Conservation Program 8, 126
scaly legs 82, *180*, 186
screen
    security camera **76–77**, 119, 193, 200
segmented worms (Annelida) 145, 165
        earthworms 70, 105, 146, **166**, 186, 219, 220
        whiteworms 105, 146, **165–166**, 216
self-feeder (live food) 13, 70, **105**, 137, 148
service corridor 54–55, 59, 67, 86, 101
sex determination (see also DNA testing) *18*, 27, **33**, 35, 82
shift box 42, 54, 56, 70
shipping 37, 41, **45**, 54, 67–68, 78, 80, 85, 88, 130, 173, 176, 192
shrew 73
singing/song (vocalization) 34–36, 92
        imprinting 36, 124
        repertoire 34
        signature song 35
Single Population Analysis Record Keeping System (see SPARKS)
sinus and eye disorders 178–179
size (body) *21*, 27, 111
skin 32, 88, 133
    lesions 69, 85, 182, 190
sleeping *22*, 31–32, 57. 75, 109, 179, 191
    boxes/cavities 211, 212
slugs (mollusc) 70, 105, 186
snake 72
sneezing 178, 187
Society for the Prevention of Cruelty to Animals (see SPCA)
sow bugs 105
SPARKS (Single Population Analysis Record Keeping System) 171
SPCA (Society for the Prevention of Cruelty to Animals) 73
species specific (traits) 29, 36, 95, 124–125, 169, 186
Species Survival Commission [of the World Conservation Union (IUCN)] (see SSC and IUCN)
spiders 105, 107, 144–145, 166
"spot-on" treatment 188
SSC (Species Survival Commission) 171
starvation **193**, 186
staying on the ground 183
stereotyped movements (stereotypy) 31, 93, 192
stool 179
    sample 44, **179**, 189, 190
stress *23*, 11, 35, 37, 39, 45, 49, 54, 67, 78, 80, 84, 87–88, 99, 102, 108–110, 132, 134, 175–176, 178-179, 184, 186–187, **191–192**
    stress-bands 184
studbooks 14, 79, 94, 114
    international 169, 171
    pekin robins 8, 113, 172
styptic agent 185
subspecies **16–***17*, 28
sunbathing 31–32
swollen feet *180–181*

tail 27, 59, 185
    bobbing 83, 178
    coverts 15, 27, 33
    loss *18*, 27, 37, 67–68, 82
tarsus 111–112
taxonomy 7, 27, 196
territory 32, 35, 61, 92
    breeding *19*, 48, 76, **96–99**, 170, 196
threadworms (see Capillaria worms and parasites)
toenails 34, **60**, 81, 182, 185
trace elements 130, **134–135**
transporting 42, 68
trapping 20, **68–70**, 72–74, 89
    cage 42, 45, 54
    Havahart 72
    insects *23*, 161–162, 167, *20*
    rodents 71–72, *20*
    toenail in leg band 111–112
trout chow 160–161

uric acid 180–181, 189

vegetation (see plants)
vent
    airflow 44–46, 53, 56, 85, 147–148, 153, 155
    bird anatomy 81–82, 179, 185
veterinarian 86, 175–183
viral infection 179

vitamins 88, 96, 106–107, 121, 132–134, 150, 159, 165, 177, 185
   vitamin A 96, 108, 132–134, 139–142, 179, 180
   vitamin B-complex 123, 139–140, 185
   vitamin D 96, 108, 133–134, 139–141, 185
   vitamin E 133, 139–141
   vitamin K 133
   vitamin C 132–133, 135, 139, 140
   pro-vitamins 132–133, 185
vocalization (see singing)

WAZA (World Association of Zoos and Aquariums) 171, 240
walls 50, 71, 75
wasps 70, 105
water 31, 61–62, 86, 149
   bath *20*, 59, 137
   drinking, birds 31, 42, 46, 48, 58, 61, 85, 87–88, 121–123, **130**, 137, 177, 180, 193
   drinking, insects 150–151, 157, 161, 163–164
   watering plants 53, 59, 63, 137

waxworms *23*, 88, 105, 107–108, 121, 127, 136, 138, 140–141, 145, 147, 176
   containers, hatching, rearing 155–156
   culture cabinets 155
   food medium 156
weaning *22*, 94, 110, 113, 122
weather exposure 193
weight loss 123, 187
whiteworms (potworms) 105, 146, **165**
wild birds 9, 13, 40, 53, 86, 176, 186
windpipe (see larynx)
wings, injured 182
World Association of Zoos and Aquariums (see WAZA)
World Conservation Union (see IUCN)

yeast infection 178

zoonotic diseases 178

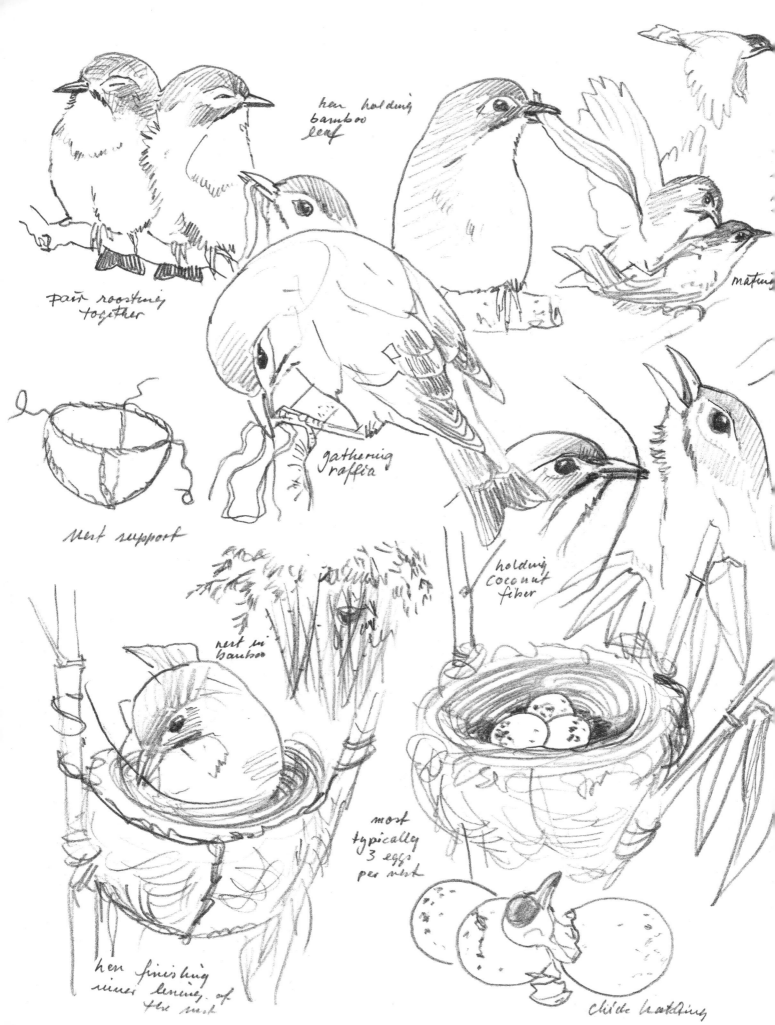